Lecture Notes in Computer Scie

T0238378

Commenced Publication in 1973
Founding and Former Series Editors:
Gerhard Goos, Juris Hartmanis, and Jan van Leeuwen

Joachim von zur Gathen
José Luis Imaña
Çetin Kaya Koç (Eds.)

Arithmetic
of Finite Fields

2nd International Workshop, WAIFI 2008
Siena, Italy, July 6–9, 2008
Proceedings

 Springer

Volume Editors

Joachim von zur Gathen
B-IT, Universität Bonn
Dahlmannstr. 2
53113 Bonn, Germany
E-mail: gathen@bit.uni-bonn.de

José Luis Imaña
Complutense University
28040 Madrid, Spain
E-mail: jluimana@dacya.ucm.es

Çetin Kaya Koç
Istanbul Chamber of Commerce
34112 Istanbul, Turkey,
E-mail: koc@cryptocode.net

Library of Congress Control Number: 2008929536

CR Subject Classification (1998): E.4, I.1, E.3, G.2, F.2

LNCS Sublibrary: SL 1 – Theoretical Computer Science and General Issues

ISSN   0302-9743
ISBN-10   3-540-69498-6 Springer Berlin Heidelberg New York
ISBN-13   978-3-540-69498-4 Springer Berlin Heidelberg New York

Springer is a part of Springer Science+Business Media

springer.com

© Springer-Verlag Berlin Heidelberg 2008
Printed in Germany

Typesetting: Camera-ready by author, data conversion by Scientific Publishing Services, Chennai, India
Printed on acid-free paper   SPIN: 12323364   06/3180   5 4 3 2 1 0

Preface

These are the proceedings of WAIFI 2008, the second workshop on the Arithmetic of Finite Fields, that was held in Siena, Italy, July 6-9, 2008. The first workshop, WAIFI 2007, which was held in Madrid (Spain), was received quite enthusiastically by mathematicians, computer scientists, engineers and physicists who are performing research on finite field arithmetic. We believe that there is a need for a workshop series bridging the gap between the mathematical theory of finite fields and their hardware/software implementations and technical applications. We hope that the WAIFI workshop series, which from now on will be held on even years, will help to fill this gap.

There were 34 submissions to WAIFI 2008, of which the Program Committee selected 16 for presentation. Each submission was reviewed by at least three reviewers. Our thanks go to the Program Committee members for their many contributions and hard work. We are also grateful to the external reviewers listed below for their expertise and assistance in the deliberations. In addition to the contributions appearing in these proceedings, the workshop program included an invited lecture given by Amin Shokrollahi.

Special compliments go out to Enrico Martinelli, General Co-chair, and to Roberto Giorgi and Sandro Bartolini, local organizers of WAIFI 2008, who brought the workshop to Siena, one of the most beautiful cities of Tuscany, Italy. WAIFI 2008 was organized by the Dipartimento di Ingegneria dell'Informazione of the University of Siena, Italy.

The submission and selection of papers were done using the iChair software, developed at EPFL by Thomas Baignères and Matthieu Finiasz. We also thank Deniz Karakoyunlu for his help in this matter.

July 2008

Joachim von zur Gathen
José Luis Imaña
Çetin Kaya Koç

Organization

Steering Committee

Claude Carlet — University of Paris 8, France
Jean-Pierre Deschamps — University Rovira i Virgili, Spain
José Luis Imaña — Complutense University of Madrid, Spain
Çetin Kaya Koç — Oregon State University, USA
Christof Paar — Ruhr University of Bochum, Germany
Jean-Jacques Quisquater — Université catholique de Louvain, Belgium
Berk Sunar — Worcester Polytechnic Institute, USA
Gustavo Sutter — Autonomous University of Madrid, Spain

Executive Committee

General Co-chairs

José Luis Imaña — Complutense University of Madrid, Spain
Enrico Martinelli — University of Siena, Italy

Program Co-chairs

Joachim von zur Gathen — B-IT, University of Bonn, Germany
Çetin Kaya Koç — Oregon State University, USA

Financial, Local Arrangements Chairs

Sandro Bartolini — University of Siena, Italy
Roberto Giorgi — University of Siena, Italy

Publicity Chair

Claude Carlet — University of Paris 8, France

Program Committee

Omran Ahmadi — University of Waterloo, Canada
Daniel Augot — INRIA-Rocquencourt, France
Jean-Claude Bajard — University of Montpellier II, France
Luca Breveglieri — Politecnico di Milano, Italy
Stephen Cohen — University of Glasgow, UK
Ricardo Dahab — Universidade Estadual de Campinas, Brazil
Gianluca Dini — University of Pisa, Italy
Serdar Erdem — Gebze Institute of Technology, Turkey
Joachim von zur Gathen — B-IT, University of Bonn, Germany

Elisa Gorla   University of Zürich, Switzerland
Dirk Hachenberger   University of Augsburg, Germany
Anwar Hasan   University of Waterloo, Canada
Marc Joye   Thomson R&D, France
Çetin Kaya Koç   Oregon State University, USA
Arjen Lenstra   EPFL, Switzerland
Peter Montgomery   Microsoft Research, USA
Ferruh Özbudak   Middle East Technical University, Turkey
Francesco Pappalardi   University of Rome 3, Italy
Francisco Rodríguez-Henríquez   Cinvestav, Mexico
René Schoof   University of Rome 2, Italy
Éric Schost   University of Western Ontario, Canada
Jamshid Shokrollahi   Ruhr University Bochum, Germany
Berk Sunar   Worcester Polytechnic Institute, USA
Chris Umans   California Institute of Technology, USA
Colin Walter   Comodo Research Lab, UK

Referees

A. Barenghi   D. Karakoyunlu   A. Reyhani-Masoleh
L. Batina   A. Karlov   M. Roetteler
A. Canteaut   S. Khazaei   G. Saldamlı
C. Carlet   C. Lauradoux   J. Sarinay
P. Charpin   D. Loebenberger   S. Sarkar
N. Courtois   M. Macchetti   E. Savas
J. Detrey   W. Marnane   O. Schütze
L. El Aimani   F. Morain   I. Shparlinski
H. Fan   C. Negre   M. Stam
S. Fischer   M. Nüsken   R. Venkatesan
F. Fontein   S. Paul   J. Zumbrägel
P. Gaborit   G. Pelosi
M. Kaihara   T. Plantard

Sponsoring Institutions

Microsoft Research.
CINECA - Inter University Computing Centre, Italy
University of Siena, Italy

Table of Contents

Codes and Cryptography

Interpolation of the Double Discrete Logarithm

Gerasimos C. Meletiou[1] and Arne Winterhof[2]

[1] A.T.E.I. of Epirus
P.O. Box 110, GR 47100,
Arta, Greece
gmelet@teiep.gr

[2] Johann Radon Institute for Computational and Applied Mathematics
Austrian Academy of Sciences
Altenbergerstr. 69, 4040 Linz, Austria
arne.winterhof@oeaw.ac.at

Abstract. The double discrete logarithm has attracted interest as a one-way function in cryptography, in particular in group signature schemes and publicly verifiable secret sharing schemes. We obtain lower bounds on the degrees of polynomials interpolating the double discrete logarithm in multiplicative subgroups of a finite field and in the group of points on an elliptic curve over a finite field, respectively. These results support the assumption of hardness of the double discrete logarithm if the parameters are properly chosen. Similar results for other cryptographic one-way functions including the discrete logarithm, the Diffie-Hellmann mapping and related functions as well as functions related to the integer factoring problem have already been known to the literature. The investigations on the double discrete logarithm in this paper are motivated by these results on other cryptographic functions.

Keywords: double discrete logarithm, interpolation polynomials, finite fields, elliptic curves.

1 Introduction

Let G be a cyclic group of order t generated by an element g. We identify the residue class ring $\mathbb{Z}_t$ of order t with the set of integers $\{0, 1, \ldots, t-1\}$. Let $h \in \mathbb{Z}_t^*$ an element of order m. For $0 \leq x < m$ the *double discrete logarithm* $\mathrm{ddl}(z)$ of an element $z = g^{h^x} \in G$ is defined as $\mathrm{ddl}(z) = x$.

The parameters G, t, g and h should be chosen such that computing discrete logarithms in G to the base g and in $\mathbb{Z}_t^*$ to the base h are infeasible.

The double discrete logarithm is used as a one-way function in several cryptographic schemes, in particular in group signature schemes and publicly verifiable secret sharing schemes, see [3, 4, 8, 9, 10, 12, 17, 19, 25, 37, 38, 39].

In this note we consider two important classes of groups G,

1. multiplicative subgroups of order t of a finite field $\mathbb{F}_q$ with q elements,
2. groups of points on elliptic curves over a finite field $\mathbb{F}_q$ generated by a point of order t.

J. von zur Gathen, J.L. Imaña, and Ç.K. Koç (Eds.): WAIFI 2008, LNCS 5130, pp. 1–10, 2008.
© Springer-Verlag Berlin Heidelberg 2008

For these two kinds of groups we show that there are no low degree interpolation polynomials of the double discrete logarithm for a large set of given data if the parameters are properly chosen. These results support the assumption of hardness of the double discrete logarithm.

The investigations of this paper are motivated by similar results on functions related to the discrete logarithm and the integer factoring problem, respectively. See the monograph [35] and the series of papers [1, 2, 6, 7, 11, 13, 14, 15, 21, 22, 23, 24, 26, 27, 28, 30, 31, 32, 33, 40, 41].

2   Subgroups of Finite Prime Fields

We start with the case where G is a subgroup of the multiplicative group of the finite field $\mathbb{F}_p$ where p is a prime with $p > 5$.

Theorem 1. *Let $t \geq 3$ be an integer, p be a prime with $p \equiv 1 \bmod t$, $g \in \mathbb{F}_p^*$ an element of order t, $h \in \mathbb{Z}_t^*$ an element of order $m \geq 2$ and $S \subseteq \{0, 1, \ldots, m-1\}$ a set of order $|S| = m - s$. Let $f(X) \in \mathbb{F}_p[X]$ be a polynomial satisfying*

$$f\left(g^{h^n}\right) = n, \quad n \in S,$$

then we have

$$\deg(f) \geq \frac{m - 2s}{2v},$$

where v is the smallest integer in the set $\{h^n \bmod t : 1 \leq n < m\}$.

Proof. Define y by $v = h^y$ and $1 \leq y < m$, and consider the set

$$R = \{n \in S : (n + y \bmod m) \in S\}.$$

Obviously we have
$$|R| \geq |S| - s = m - 2s.$$

For $n \in R$ we have

$$f(g^{vh^n}) = f(g^{h^{n+y}}) = n + y + \delta = f(g^{h^n}) + y + \delta$$

with $\delta \in \{0, -m\}$. Hence, one of the two polynomials

$$F_\delta(X) = f(X^v) - f(X) - y - \delta, \quad \delta \in \{0, -m\},$$

of degree $\deg(F) = v \deg(f)$ has at least $|R|/2$ zeros and we get

$$\deg(f) = \frac{\deg(F)}{v} \geq \frac{|R|}{2v} \geq \frac{m - 2s}{2v},$$

which completes the proof. $\square$

Remark. In the probably most important case when t is a prime, see e.g. [37, 38], and m is large, i.e. $m = (t-1)/d$ with a small d, we have $v \leq 2^d$. For the case

$d = 2$ of [37, 38], v is the smallest quadratic residue modulo t larger than 1, i.e. $v = 2$ if $t \equiv \pm 1 \bmod 8$, $v = 3$ if $t \equiv \pm 11 \bmod 24$ and $v = 4$ if $t \equiv \pm 3$ or $\pm 5 \bmod 24$ by the quadratic reciprocity law and its supplement. In general we trivially have $v \leq h$. Moreover, unless m is very small we have $v = o(m)$, see [16]. For example, if t is prime and $m \geq t^{1/2}$ then $v = O\left(t^{34/37+\varepsilon}\right)$, see [18, Theorem 7.10].

3   Subgroups of Arbitrary Finite Fields

Now let $q = p^r \equiv 1 \bmod t$ be a power of a prime p. For $m \leq p$ the bound of Theorem 1 is still valid. However, for $m > p$ we lose information since we consider the double discrete logarithm modulo p. A compensation for the interpolation polynomial of the double discrete logarithm is the function defined in the sequel.

Let $\{\beta_1, \ldots, \beta_r\}$ be a basis of $\mathbb{F}_q$ over $\mathbb{F}_p$. Then for $n \geq 0$ we define $\xi_n \in \mathbb{F}_q$ by

$$\xi_n = n_1 \beta_1 + \ldots + n_r \beta_r$$

if

$$n \equiv n_1 + \ldots + n_r p^{r-1} \bmod q, \quad 0 \leq n_1, \ldots, n_r < p.$$

Then instead of the double discrete logarithm we consider the mapping

$$\mathrm{ddl}^*(g^{h^n}) \mapsto \xi_n, \quad 0 \leq n < m.$$

Theorem 2. *Let $t \geq 3$ be an integer, p be a prime and r an integer such that $q = p^r \equiv 1 \bmod t$, $g \in \mathbb{F}_q^*$ an element of order t, $h \in \mathbb{Z}_t^*$ an element of order $m \geq 2$ and $S \subseteq \{0, 1, \ldots, m-1\}$ a set of order $|S| = m - s$. Let $f(X) \in \mathbb{F}_q[X]$ be a polynomial satisfying*

$$f\left(g^{h^n}\right) = \xi_n, \quad n \in S,$$

then we have

$$\deg(f) \geq \max\left\{ \frac{m - 2s}{2^l v}, \frac{m - 4s}{2h} \right\},$$

where $l = \lceil \log_p(m) \rceil$ and v is the smallest integer in the set $\{h^n \bmod t : 1 \leq n < m\}$.

Proof. First we proceed as in the proof of Theorem 1 and use the same notation. For $n \in R$ we have

$$f(g^{vh^n}) = \xi_{n+y+\delta} = \xi_n + \omega = f(g^{h^n}) + \omega$$

for at most 2^l different elements $\omega \in \mathbb{F}_q$, where $\delta = 0$ if $n + y < m$ and $\delta = -m$ otherwise.

More precisely, if $\delta = 0$ then we have $n + y < m < q$. Let

$$n = n_0 + n_1 p + \ldots + n_{l-1} p^{l-1}, \quad y = y_0 + y_1 p + \ldots + y_{l-1} p^{l-1}$$

and

$$n + y = z_0 + z_1 p + \ldots + z_{l-1} p^{l-1}$$

with $0 \le n_0, n_1, \ldots, n_{l-1}, y_0, y_1, \ldots, y_{l-1}, z_0, z_1, \ldots, z_{l-1} < p$ be the p-adic expansions of n, y and $n + y$. Put $k_0 = k_l = 0$ and

$$k_j = \begin{cases} 0 \text{ if } n_{j-1} + y_{j-1} + k_{j-1} < p, \\ 1 \text{ otherwise,} \end{cases} \quad \text{where } j = 1, 2, \ldots, l-1.$$

Then we have

$$z_j = n_j + y_j + k_j - k_{j+1} p, \quad j = 0, \ldots, l-1.$$

Hence, we have $\omega = \xi_y + \xi_k$ where $k = k_1 p + \ldots + k_{l-1} p^{l-1}$ and there are at most 2^{l-1} possible ω.

If $\delta = -m$ we get similarly

$$\xi_n = \xi_{(n+y-m)+(m-y)} = \xi_{n+y-m} + \xi_{m-y} + \xi_k$$

for at most 2^{l-1} different k and thus $\xi_{n+y+\delta} = \xi_n + \omega$ with $\omega = -\xi_{m-y} - \xi_k$.

Hence at least one of the 2^l polynomials $F_\omega(X) = f(X^v) - f(X) - \omega$ has at least $|R|/2^l$ zeros and thus $v \deg(f) = \deg(F_\omega) \ge |R|/2^l$ and the first result follows.

Similarly, we see that

$$f(g^{h^{n+1}}) = \xi_{n+1} = \xi_n + \xi_1 = f(g^{h^n}) + \xi_1$$

for all $n \in S$ with $n + 1 \in S$ and $n \not\equiv -1 \bmod p$. Hence, the polynomial

$$F(X) = f(X^h) - f(X) - 1$$

of degree equal to $h \deg(f)$ has at least

$$\left(1 - \frac{1}{p}\right) m - 2s \ge \frac{m}{2} - 2s$$

zeros and the second result follows. □

Remark. Note that the first bound $\deg(f) \ge (m - 2s)/(2^l v)$ is only strong if p is large and trivial for $p = 2$. In this particularly interesting case only the second bound $\deg(f) \ge (m - 4s)/(2h)$ applies.

4 Elliptic Curves

Let E be an elliptic curve over the finite field $\mathbb{F}_q$ defined by the Weierstraß equation

$$E : Y^2 + h(X)Y = f(X)$$

with a linear polynomial

$$h(X) = a_1X + a_3, \quad a_1, a_3 \in \mathbb{F}_q,$$

and a cubic polynomial

$$f(X) = X^3 + a_2X^2 + a_4X + a_6, \quad a_2, a_4, a_6 \in \mathbb{F}_q,$$

such that over the algebraic closure $\overline{\mathbb{F}_q}$ there are no solutions $(x, y) \in \overline{\mathbb{F}_q}^2$ simultaneously satisfying the equations

$$y^2 + h(x)y = f(x), \quad 2y + h(x) = 0, \quad \text{and} \quad h'(x)y = f'(x).$$

In odd characteristic we may assume $Y^2 = f(X)$ and in even characteristic $h(X) = X$ or $h(X) = 1$. The latter case corresponds to *supersingular curves*. We denote by $\mathcal{O}$ the point at infinity.

We restrict ourselves to the two important cases where $q = p > 3$ is a prime and where $q = 2^r$ and the curve is not supersingular.

4.1   Elliptic Curves over Fields of Prime Order

For an elliptic curve over $\mathbb{F}_p$ with $p > 3$ we can assume that E is defined by an equation of the form

$$E : Y^2 = X^3 + aX + b, \quad a, b \in \mathbb{F}_p.$$

We recall some basic facts on division polynomials (see e. g. [5, 20, 29, 34, 36]). The *division polynomials* $\psi_v(X, Y) \in \mathbb{F}_p[X, Y]/(Y^2 - X^3 - aX - b)$, $v \geq 0$, are recursively defined by

$$\psi_0 = 0,$$
$$\psi_1 = 1,$$
$$\psi_2 = 2Y,$$
$$\psi_3 = 3X^4 + 6aX^2 + 12bX - a^2,$$
$$\psi_4 = 4Y(X^6 + 5aX^4 + 20bX^3 - 5a^2X^2 - 4abX - 8b^2 - a^3),$$
$$\psi_{2k+1} = \psi_{k+2}\psi_k^3 - \psi_{k+1}^3\psi_{k-1}, \quad k \geq 2,$$
$$\psi_{2k} = \psi_k(\psi_{k+2}\psi_{k-1}^2 - \psi_{k-2}\psi_{k+1}^2)/(2Y), \quad k \geq 3,$$

where ψ_v is an abbreviation for $\psi_v(X, Y)$. If v is odd then $\psi_v(X, Y) \in \mathbb{F}_p[X]$ is univariate and $\psi_v(X, Y) \in Y\mathbb{F}_p[X]$ if v is even. Therefore, as $Y^2 = X^3 + aX + b$, we have $\psi_v^2(X, Y), \psi_{v-1}(X, Y)\psi_{v+1}(X, Y) \in \mathbb{F}_p[X]$. In particular, we may write $\psi_{v-1}\psi_{v+1}(X)$ and $\psi_v^2(X)$.

The division polynomials can be used to determine multiples of a point. Let $P = (x, y) \neq \mathcal{O}$, then the first coordinate of vP is given by

$$\frac{\theta_v(x)}{\psi_v^2(x)}, \quad \text{where } \theta_v(X) = X\psi_v^2(X) - \psi_{v-1}\psi_{v+1}(X).$$

The zeros of the denominator $\psi_v^2(X)$ are exactly the first coordinates of the nontrivial v-torsion points, i.e., the points $Q = (x, y) \in \overline{\mathbb{F}_p}^2$ on E with $vQ = \mathcal{O}$. The x-coordinates of a v-torsion point Q cannot be a zero of $\theta_v(X)$ since otherwise the x-coordinate of vQ would be 0 in contradiction to $vQ = \mathcal{O}$.

We recall that the degree of $\psi_v^2(X)$ is $v^2 - 1$ if $p \nmid v$. The polynomial $\theta_v(X) \in \mathbb{F}_p[X]$ is monic of degree v^2.

Theorem 3. *Let $p > 3$ be a prime, E an elliptic curve over $\mathbb{F}_p$ and P a point on E of prime order t. For $1 \le k \le t - 1$ let the first coordinate of kP be denoted by x_k. Let $h \in \mathbb{Z}_t^*$ be an element of order $m \ge 2$ and let S be a subset of $\{0, 1, \ldots, \min\{m, p\} - 1\}$ of cardinality $m - s$. Let $F(X) \in \mathbb{F}_p[X]$ satisfy*

$$F(x_{h^n}) = n, \quad n \in S,$$

then we have

$$\deg(F) \ge \frac{1}{4 \cdot 2^{2(t-1)/m}} \left(\min\{m, p\} - 2s\right).$$

Proof. Note that $x_k = x_{k'}$ if and only if $k \equiv \pm k' \bmod t$. (There are at most two points on E with the same first coordinate. These two points add to $\mathcal{O}$.) Hence, if m is odd all x_{h^n} for $0 \le n \le m - 1$ are different and if m is even exactly two elements $x_{h^n} = x_{h^{n+m/2}}$ coincide for $0 \le n \le m/2 - 1$.

Obviously, for $v \equiv 2^{(t-1)/m} \bmod t$ the first coordinate x_v of vP has a unique representation with $v = h^{n_0}$ and $1 \le n_0 \le m-1$ if m is odd or $1 \le n_0 \le m/2-1$ if m is even. The subset R of S defined by

$$R = \{n \in S : (n + n_0 \bmod m) \in S\}$$

has cardinality at least

$$|R| \ge |S| - s = \min\{m, p\} - 2s.$$

Put $d = \deg(F)$. Using the polynomials $\psi_v^2(X)$ and $\theta_v(X)$ defined above we get for $n \in R$,

$$F\left(\frac{\theta_v(x_{h^n})}{\psi_v^2(x_{h^n})}\right) = F(x_{vh^n}) = n + n_0 + \delta = F(x_{h^n}) + n_0 + \delta,$$

where $\delta \in \{0, -m\}$. Finally, we consider the polynomials

$$U_\delta(X) = \psi_v^{2d}(X)\left(F\left(\frac{\theta_v(X)}{\psi_v^2(X)}\right) - F(X) - n_0 - \delta\right).$$

Let $\alpha \in \overline{\mathbb{F}_p}$ be a zero of $\psi_v^2(X)$ and thus not a zero of $\theta_v(X)$. Then we have

$$U_\delta(\alpha) = a_d \theta_v(\alpha) \ne 0,$$

where a_d is the leading coefficient of $F(X)$. Thus $U_\delta(X)$ is not identical to zero and has $\deg(U_\delta) \le v^2 d$. At least one of the polynomials $U_\delta(X)$ has at least $|R|/4$ zeros and we get $d \ge \deg(U_\delta)/v^2 \ge |R|/(4v^2)$ and thus the result. $\qquad\square$

Remark. Note that by the Hasse-Weil Theorem we have $m < t \le p + 1 + 2p^{1/2}$ and the only case when $m > p$ is possible is $m = t - 1 \ge p$. However, even in this case only at most $O(m^{1/2})$ function values of the double discrete logarithm are not considered in the theorem.

4.2 Non-supersingular Elliptic Curves over Finite Fields of Characteristic 2

As in Section 3 we have to deal with $\mathbb{F}_{2^r}$ in a different way.

We consider a non-supersingular elliptic curve over $\mathbb{F}_{2^r}$ defined by an equation of the form

$$Y^2 + XY = X^3 + aX^2 + b.$$

The division polynomials $\psi_v(X) \in \mathbb{F}_{2^r}[X]$ are defined by

$$\psi_0 = 0,$$
$$\psi_1 = 1,$$
$$\psi_2 = X,$$
$$\psi_3 = X^4 + X^3 + b,$$
$$\psi_4 = X^6 + bX^2,$$
$$\psi_{2k+1} = \psi_{k+2}\psi_k^3 + \psi_{k+1}^3\psi_{k-1}, \quad k \ge 2,$$
$$\psi_{2k} = \psi_k(\psi_{k+2}\psi_{k-1}^2 + \psi_{k-2}\psi_{k+1}^2)/X, \quad k \ge 3.$$

The degree of ψ_v is obviously $(v^2 - 1)/2$ if v is odd and at most $(v^2 - 2)/2$ if v is even. For $P = (x, y) \ne \mathcal{O}$, the first coordinate of vP is given by

$$x + \frac{\psi_{v-1}(x)\psi_{v+1}(x)}{\psi_v(x)^2}.$$

Theorem 4. *Let E be a non-supersingular curve over $\mathbb{F}_{2^r}$ and P a point on E of order t. For $1 \le k \le t - 1$ let the first coordinate of kP be denoted by x_k. Let $h \in \mathbb{Z}_t^*$ be an element of order $m \ge 2$ and let S be a subset of $\{0, 1, \ldots, m - 1\}$ of cardinality $m - s$. Let $F(X) \in \mathbb{F}_p[X]$ satisfy*

$$F(x_{h^n}) = \xi_n, \quad n \in S,$$

then we have

$$\deg(F) \ge \frac{m - 4s}{2h^2}.$$

Proof. We proceed as in the previous proofs and use the same notation. We have

$$F\left(x_{h^n} + \frac{\psi_{h-1}(x_{h^n})\psi_{h+1}(x_{h^n})}{\psi_h(x_{h^n})^2}\right) = F(x_{h^{n+1}}) = \xi_{n+1} = \xi_n + \xi_1$$

for all $n \in S$ with $n + 1 \in S$ and $n \equiv 0 \bmod 2$. Put $d = \deg(F)$. Hence the polynomial

$$U(X) = \psi_h(X)^{2d}\left(F\left(X + \frac{\psi_{h-1}(X)\psi_{h+1}(X)}{\psi_h(X)^2}\right) - F(X) - \xi_1\right)$$

of degree $h^2 d$ has at least $m/2 - 2s$ zeros and the result follows. □

Acknowledgments

The second author was supported by the Austrian Science Fund FWF under research grant P19004-N18. The paper was partially written during a visit of the first author to RICAM. He wishes to express his gratitude to the Austrian Academy of Sciences for the hospitality.

References

1. Adelmann, C., Winterhof, A.: Interpolation of functions related to the integer factoring problem. In: Ytrehus, Ø. (ed.) WCC 2005. LNCS, vol. 3969, pp. 144–154. Springer, Heidelberg (2006)
2. Aly, H., Winterhof, A.: Polynomial representations of the Lucas logarithm. Finite Fields Appl. 12(3), 413–424 (2006)
3. Ateniese, G., Tsudik, G.: Some open issues and new directions in group signatures. In: Franklin, M. (ed.) FCT 1999. LNCS, vol. 1684, pp. 196–211. Springer, Heidelberg (1999)
4. Ateniese, G., Song, D., Tsudik, G.: Quasi efficient revocation group signatures. In: Blaze, M. (ed.) FC 2002. LNCS, vol. 2357, pp. 183–197. Springer, Heidelberg (2003)
5. Blake, I.F., Seroussi, G., Smart, N.: Elliptic Curves in Cryptography. Reprint of the 1999 original. London Mathematical Society Lecture Note Series, vol. 265. Cambridge University Press, Cambridge (2000)
6. Brandstätter, N., Lange, T., Winterhof, A.: On the non-linearity and sparsity of Boolean functions related to the discrete logarithm in finite fields of characteristic two. In: Ytrehus, Ø. (ed.) WCC 2005. LNCS, vol. 3969, pp. 135–143. Springer, Heidelberg (2006)
7. Brandstätter, N., Winterhof, A.: Approximation of the discrete logarithm in finite fields of even characteristic by real polynomials. Arch. Math. (Brno) 42(1), 43–50 (2006)
8. Bussard, L., Roudier, Y., Molva, R.: Untraceable secret credentials: trust establishment with privacy. Pervasive Computing and Communications Workshops, 2004. In: Proceedings of the Second IEEE Annual Conference, March 14-17, 2004, pp. 122–126 (2004)
9. Camenisch, J.: Group signature schemes and payment systems based on the discrete logarithm problem. Phd-thesis, ETH Zürich, Diss. ETH No. 12520 (1998)
10. Camenisch, J., Stadler, M.: Efficient group signature schemes for large groups. In: Kaliski Jr., B.S. (ed.) CRYPTO 1997. LNCS, vol. 1294, pp. 410–424. Springer, Heidelberg (1997)
11. Coppersmith, D., Shparlinski, I.: On polynomial approximation of the discrete logarithm and the Diffie-Hellman mapping. J. Cryptology 13(3), 339–360 (2000)
12. Elkamchouchi, H.M., Nasr, M.E., Esmail, R.: New public key techniques based on double discrete logarithm problem. Radio Science Conference, 2004. NRSC 2004. In: Proceedings of the Twenty-First National, vol. C23, pp. 1–9 (2004)
13. El Mahassni, E., Shparlinski, I.E.: Polynomial representations of the Diffie-Hellman mapping. Bull. Austral. Math. Soc. 63, 467–473 (2001)
14. Kiltz, E., Winterhof, A.: Lower bounds on weight and degree of bivariate polynomials related to the Diffie-Hellman mapping. Bull. Austral. Math. Soc. 69, 305–315 (2004)

15. Kiltz, E., Winterhof, A.: Polynomial interpolation of cryptographic functions related to Diffie-Hellman and discrete logarithm problem. Discrete Appl. Math. 154, 326–336 (2006)
16. Korobov, N.M.: The distribution of digits in periodic fractions. Mat. Sb. (N.S.) 89(131), 654–670, 672 (1972) (Russian)
17. Konoma, C., Mambo, M., Shizuya, H.: The computational difficulty of solving cryptographic primitive problems related to the discrete logarithm problem. IEICE Transactions on Fundamentals of Electronics, Communications and Computer Sciences E88-A(1), 81–88 (2005)
18. Konyagin, S.V., Shparlinski, I.E.: Character Sums with Exponential Functions and Their Applications. Cambridge Tracts in Mathematics, vol. 136. Cambridge University Press, Cambridge (1999)
19. Kula, M.A.: A cryptosystem based on double exponentiation. Tatra Mt. Math. Publ. 25, 67–80 (2002)
20. Lang, S.: Elliptic Curves: Diophantine Analysis. Springer, Berlin (1978)
21. Lange, T., Winterhof, A.: Polynomial Interpolation of the Elliptic Curve and XTR Discrete Logarithm. In: Ibarra, H.O., Zhang, L. (eds.) COCOON 2002. LNCS, vol. 2387, pp. 137–143. Springer, Heidelberg (2002)
22. Lange, T., Winterhof, A.: Incomplete character sums over finite fields and their application to the interpolation of the discrete logarithm by Boolean functions. Acta Arith. 101, 223–229 (2002)
23. Lange, T., Winterhof, A.: Interpolation of the discrete logarithm in F_q by Boolean functions and by polynomials in several variables modulo a divisor of $q - 1$. In: International Workshop on Coding and Cryptography (WCC 2001), Paris (2001); Discrete Appl. Math. 128, 193–206 (2003)
24. Lange, T., Winterhof, A.: Interpolation of the elliptic curve Diffie-Hellman mapping. In: Fossorier, M.P.C., Høholdt, T., Poli, A. (eds.) AAECC 2003. LNCS, vol. 2643, pp. 51–60. Springer, Heidelberg (2003)
25. Lysyanskaya, A., Ramzan, Z.: Group blind digital signatures: A scalable solution to electronic cash. In: Hirschfeld, R. (ed.) FC 1998. LNCS, vol. 1465, pp. 184–197. Springer, Heidelberg (1998)
26. Meidl, W., Winterhof, A.: A polynomial representation of the Diffie-Hellman mapping. Appl. Algebra Engrg. Comm. Comput. 13, 313–318 (2002)
27. Meletiou, G.C.: Explicit form for the discrete logarithm over the field GF(p, k). Arch. Math. (Brno) 29, 25–28 (1993)
28. Meletiou, G.C., Mullen, G.L.: A note on discrete logarithms in finite fields. Appl. Algebra Engrg. Comm. Comput. 3(1), 75–78 (1992)
29. Menezes, A.: Elliptic curve public key cryptosystems. Communications and Information Theory. The Kluwer International Series in Engineering and Computer Science, vol. 234. Kluwer Academic Publishers, Boston (1993)
30. Mullen, G.L., White, D.: A polynomial representation for logarithms in GF(q). Acta Arith. 47(3), 255–261 (1986)
31. Niederreiter, H.: A short proof for explicit formulas for discrete logarithms in finite fields. Appl. Algebra Engrg. Comm. Comput. 1(1), 55–57 (1990)
32. Niederreiter, H., Winterhof, A.: Incomplete character sums and polynomial interpolation of the discrete logarithm. Finite Fields Appl. 8(2), 184–192 (2002)
33. Satoh, T.: On degree of polynomial interpolations related to elliptic curve cryptography. In: Ytrehus, Ø. (ed.) WCC 2005. LNCS, vol. 3969, pp. 155–163. Springer, Heidelberg (2006)
34. Schoof, R.: Elliptic curves over finite fields and the computation of square roots mod. p. Math. Comp. 44, 483–494 (1985)

35. Shparlinski, I.E.: Cryptographic Applications of Analytic Number Theory. Complexity Lower Bounds and Pseudorandomness. Progress in Computer Science and Applied Logic, vol. 22. Birkhäuser Verlag, Basel (2003)
36. Silverman, J.H.: The Arithmetic of Elliptic Curves. Graduate texts in mathematics, vol. 106. Springer, Heidelberg (1986)
37. Stadler, M.: Publicly verifiable secret sharing. In: Maurer, U.M. (ed.) EUROCRYPT 1996. LNCS, vol. 1070, pp. 190–199. Springer, Heidelberg (1996)
38. Tso, R., Okamoto, T., Okamoto, E.: Practical strong designated verifer signature schemes based on double discrete logarithms. In: Feng, D., Lin, D., Yung, M. (eds.) CISC 2005. LNCS, vol. 3822, pp. 113–127. Springer, Heidelberg (2005)
39. Wang, G., Qing, S.: Security flaws in several group signatures proposed by Popescu. In: Gervasi, O., Gavrilova, M.L., Kumar, V., Laganá, A., Lee, H.P., Mun, Y., Taniar, D., Tan, C.J.K. (eds.) ICCSA 2005. LNCS, vol. 3482, pp. 711–718. Springer, Heidelberg (2005)
40. Winterhof, A.: A note on the interpolation of the Diffie-Hellman mapping. Bull. Austral. Math. Soc. 64, 475–477 (2001)
41. Winterhof, A.: Polynomial interpolation of the discrete logarithm. Des. Codes Cryptogr. 25, 63–72 (2002)

Finite Dedekind Sums

Yoshinori Hamahata[*]

Department of Mathematics
Tokyo University of Science, Noda, Chiba, 278-8510, Japan
hamahata_yoshinori@ma.noda.tus.ac.jp

Abstract. In this paper, we introduce Dedekind sums associated to lattices defined over finite fields. We establish the reciprocity law for them.

Keywords: Dedekind sums, lattices, Drinfeld modules.

1 Introduction

This paper is concerned with Dedekind sums in finite characteristic.

Let $c > 0, a$ be relatively prime rational integers. The classical Dedekind sum is defined as

$$s(a, c) = \sum_{k=1}^{c-1} \frac{a}{c} \left(\left\langle \frac{ak}{c} \right\rangle - \frac{1}{2} \right),$$

where $\langle x \rangle$ is a real number such that $x - \langle x \rangle \in \mathbb{Z}$ and $0 \leq \langle x \rangle < 1$. One knows some basic properties for $s(a, c)$:

(1) $s(-a, c) = -s(a, c)$.
(2) If $a \equiv a' \pmod{c}$, then $s(a, c) = s(a', c)$.
(3)(Reciprocity law) If $a, c > 0$ are coprime, then

$$s(a, c) + s(c, a) = \frac{a^2 + c^2 - 3ac + 1}{12ac}.$$

The Dedekind sum $s(a, c)$ can be written as

$$s(a, c) = \frac{1}{4c} \sum_{k=1}^{c-1} \cot \frac{k}{c} \pi \cot \frac{ak}{c} \pi.$$

Sczech [8] established analogue of Dedekind sums with elliptic functions by replacing $\cot \pi x$ by elliptic functions. In 1989, Okada [6] introduced Dedekind sums for function fields. His idea comes from Sczech's result. In function fields, there exist exponential functions attached to lattices, which are related to Drinfeld modules. These exponential functions are similar to both cotangent and elliptic functions. In particular, the exponential function related to the Carlitz module

---

[*] Partially supported by Grant-in-Aid for Scientific Research (No. 18540050), Japan Society for the Promotion of Science.

J. von zur Gathen, J.L. Imaña, and Ç.K. Koç (Eds.): WAIFI 2008, LNCS 5130, pp. 11–18, 2008.
© Springer-Verlag Berlin Heidelberg 2008

(this is a rank one Drinfeld module) play the same role as the classical exponential function. From this observation, he replaced cotangent functions by that exponential to define Dedekind sums for function fields. In [6], he established the reciprocity law for those Dedekind sums. Inspired by Okada's result, we introduce in [5] Dedekind sums for finite fields. In this exposition, we generalize Dedekind sums defined in the previous paper [5] to introduce Dedekind sums associated to lattices defined over finite fields. The main result is the reciprocity law for those Dedekind sums. The plan of the paper is as follows. In section two, we give some facts needed later. In section three, we introduce Dedekind sums associated to lattices. In section four, we present the reciprocity law for our Dedekind sums. This result will be proved in the last section.

Notations
$K = \mathbb{F}_q$: the finite field with q elements
$\overline{K}$: a fixed algebraic closure of K
$\sum'$: the sum over non-zero elements
$\prod'$: the product over non-zero elements

2   Lattices

In this section we gather some results on lattices needed later.

A *lattice* Λ in $\overline{K}$ means a linear K-subspace in $\overline{K}$ of finite dimension. For such a lattice Λ, we define the Euler product

$$e_\Lambda(z) = z \prod_{\lambda \in \Lambda}' \left(1 - \frac{z}{\lambda}\right).$$

The product defines a map $e_\Lambda : \overline{K} \to \overline{K}$. The map e_Λ has the following properties:

- e_Λ is K-linear and Λ-periodic.
- If $\dim_K \Lambda = r$, then $e_\Lambda(z)$ has the form

$$e_\Lambda(z) = \sum_{i=0}^{r} \alpha_i(\Lambda) z^{q^i}, \tag{1}$$

where $\alpha_0(\Lambda) = 1$, $\alpha_r(\Lambda) \neq 0$.
- e_Λ has simple zeros at the points of Λ, and no other zeros.
- $de_\Lambda(z)/dz = e'_\Lambda(z) = 1$. Hence we have

$$\frac{1}{e_\Lambda(z)} = \frac{e'_\Lambda(z)}{e_\Lambda(z)} = \sum_{\lambda \in \Lambda} \frac{1}{z - \lambda}. \tag{2}$$

We recall Newton formula for power sums of the zeros of a polynomial.

Proposition 1 (Newton formula cf. [3]). *Let*

$$f(X) = X^n + c_1 X^{n-1} + \cdots + c_{n-1} X + c_n$$

be a polynomial, and $\alpha_1, \ldots, \alpha_n$ the roots of $f(X)$. For each nonnegative integer k, put

$$T_k = \alpha_1^k + \cdots + \alpha_n^k.$$

Then

$$T_k + c_1 T_{k-1} + \cdots + c_{k-1} T_1 + k c_k = 0 \quad (k \le n),$$
$$T_k + c_1 T_{k-1} + \cdots + c_{n-1} T_{k-n+1} + c_n T_{k-n} = 0 \quad (k \ge n).$$

Using this formula, we have

Proposition 2. *Let Λ be a lattice in $\overline{K}$, and take a non-zero element $a \in \overline{K}$. For $m = 1, 2, \ldots, q - 2$, we have*

$$\frac{a^m}{e_\Lambda(az)^m} = \sum_{x \in \Lambda} \frac{1}{(z - x/a)^m}.$$

Proof. Let $\dim_K \Lambda = r$. We prove it by induction on m. First let us prove the case $m = 1$. The set of the roots of $e_\Lambda(az)$ is $\{x/a \mid x \in \Lambda\}$. By (2), we have

$$\frac{a}{e_\Lambda(az)} = \sum_{x \in \Lambda} \frac{1}{z - x/a}.$$

Suppose that the claim holds for $m - 1$. The polynomial $e_\Lambda(a(z - 1/X))X^{q^r}$ in the variable X has $\{1/(z - x/a) \mid x \in \Lambda\}$ as its roots. We can easily see that

$$e_\Lambda(a(z - 1/X))X^{q^r} = e_\Lambda(az)X^{q^r} - \sum_{i=0}^{r} a^i \alpha_i(\Lambda) X^{q^r - q^i}.$$

Put $T_m = \sum_{x \in \Lambda} 1/(z - x/a)^m$. Applying Newton formula to

$$e_\Lambda(az)^{-1} e_\Lambda(a(z - 1/X)) X^{q^r} = X^{q^r} - \sum_{i=0}^{r} e_\Lambda(az)^{-1} a^i \alpha_i(\Lambda) X^{q^r - q^i},$$

we get

$$T_m - \frac{a}{e_\Lambda(az)} T_{m-1} = 0$$

for $m = 1, \ldots, q - 2$. Therefore the claim holds for m. □

For $b \in \overline{K} - \{0\}$, we set

$$R(b) = \{\lambda/b \mid \lambda \in \Lambda\} - \{0\}.$$

Lemma 1

$$\sum_{x \in R(b)} x^{-m} = \begin{cases} 0 & (m = 1, \ldots, q - 2) \\ \alpha_1(\Lambda) b^{q-1} & (m = q - 1) \end{cases},$$

where $\alpha_1(\Lambda)$ is as in (1).

Proof. The set $R(b)$ consists of the non-zero roots of $e_\Lambda(bz)$. Hence $\{1/x \mid x \in R(b)\}$ is the set of all roots of

$$b^{-1}e_\Lambda(bz^{-1})z^{q^r} = \sum_{i=0}^{r} \alpha_i(\Lambda)b^{q^i-1}z^{q^r-q^i},$$

$$= z^{q^r} + \alpha_1(\Lambda)b^{q-1}z^{q^r-q} + \cdots.$$

Applying Newton formula to this polynomial, we have

$$T_m = 0 \quad (m = 1, \ldots, q-2),$$
$$T_{q-1} + (q-1)\alpha_1(\Lambda)b^{q-1} = 0.$$

The last equation yields $T_{q-1} = \alpha_1(\Lambda)b^{q-1}$. □

3   Finite Dedekind Sums

For a lattice Λ in $\overline{K}$, we define Dedekind sum.

Definition 1. Set $\widetilde{\Lambda} = \{x \in \overline{K} \mid x\lambda \in \Lambda \text{ for some } \lambda \in \Lambda\}$. *We choose* $c, a \in \overline{K} - \{0\}$ *such that* $a/c \notin \widetilde{\Lambda}$. *For* $m = 1, \ldots, q-2$, *put*

$$s_m(a,c)_\Lambda = \frac{1}{c^m} {\sum_{\lambda \in \Lambda}}' \left(\frac{\lambda}{c}\right)^{-q+1+m} e_\Lambda\left(\frac{a\lambda}{c}\right)^{-m}.$$

Moreover, we define

$$s_0(c)_\Lambda = s_0(a,c)_\Lambda = {\sum_{\lambda \in \Lambda}}' \left(\frac{\lambda}{c}\right)^{-q+1}.$$

We call $s_m(a,c)_\Lambda$ *the* m*-th Dedekind sum for* Λ.

Remark 1. In [5], we defined the Dedekind sum for $\Lambda = K$. Our definition generalizes it.

It follows from Lemma 1 that

$$s_0(c)_\Lambda = s_0(a,c)_\Lambda = \alpha_1(\Lambda)c^{q-1},$$

where $\alpha_1(\Lambda)$ is the coefficient of z^q in $e_\Lambda(z)$ as in (1).

The following result is analogous to the properties (1), (2) of the classical Dedekind sums in section one.

Proposition 3. *Dedekind sums* $s_m(a,c)_\Lambda$ $(m = 1, \ldots, q-1)$ *satisfy the following properties:*
(1) *For any* $\alpha \in K^*$, $s_m(\alpha a, c)_\Lambda = \alpha^{-m} s_m(a,c)_\Lambda$.
(2) *If* $a, a' \in \overline{K}$ *satisfy* $a - a' \in c\Lambda$, *then* $s_m(a,c)_\Lambda = s_m(a',c)_\Lambda$.

Proof. (1), (2) Immediate from the properties of $e_\Lambda(z)$. □

4   Reciprocity Law

In this section we present the reciprocity law for our Dedekind sums. Let a, c be the elements of $\overline{K} - \{0\}$ such that $a/c \notin \widetilde{\Lambda}$.

Theorem 1 (Reciprocity law I). *For $m = 1, \ldots, q - 2$, we have*

$$s_m(a, c)_\Lambda + (-1)^{m-1} s_m(c, a)_\Lambda = \sum_{r=1}^{m-1} \frac{(-1)^{m-r} s_{m-r}(c, a)_\Lambda}{a^r c^r} \cdot \binom{m+1}{r}$$
$$+ \frac{s_0(c)_\Lambda + m \cdot s_0(a)_\Lambda}{a^m c^m}.$$

We will prove it in the next section. As a corollary to this result, the next theorem is obtained.

Theorem 2 (Reciprocity law II). *For $m = 1, \ldots, q - 2$, we have*

$$s_m(a, c)_\Lambda + (-1)^{m-1} s_m(c, a)_\Lambda$$
$$= \frac{1}{2} \left\{ \sum_{r=1}^{m-1} \frac{(-1)^{r-1} \left(s_{m-r}(a, c)_\Lambda + (-1)^{m-1} s_{m-r}(c, a)_\Lambda \right) \binom{m+1}{r}}{a^r c^r} \right.$$
$$\left. + \frac{\left(m + (-1)^{m-1} \right) \left(s_0(a)_\Lambda + (-1)^{m-1} s_0(c)_\Lambda \right)}{a^m c^m} \right\}.$$

Proof. By Theorem 1,

$$s_m(a, c)_\Lambda + (-1)^{m-1} s_m(c, a)_\Lambda + (-1)^{m-1} \left(s_m(c, a)_\Lambda + (-1)^{m-1} s_m(a, c)_\Lambda \right)$$

is two times of the right hand side of the equation of the claim.   □

Example 1. Using the notation in the previous section, we have

$$s_1(a, c)_\Lambda + s_1(c, a)_\Lambda = \frac{\alpha_1(\Lambda) \left(a^{q-1} + c^{q-1} \right)}{ac},$$
$$s_3(a, c)_\Lambda + s_3(c, a)_\Lambda = \frac{2 s_2(a, c)_\Lambda + 2 s_2(c, a)_\Lambda}{ac} - \frac{\alpha_1(\Lambda) \left(a^{q-1} + c^{q-1} \right)}{a^3 c^3}.$$

In particular, if $\Lambda = K$, then $e_K(z) = z - z^q$, so that

$$s_1(a, c)_K + s_1(c, a)_K = -\frac{a^{q-1} + c^{q-1}}{ac},$$
$$s_3(a, c)_K + s_3(c, a)_K = \frac{2 s_2(a, c)_K + 2 s_2(c, a)_K}{ac} + \frac{a^{q-1} + c^{q-1}}{a^3 c^3}.$$

5   Proof of Theorem 1

We need two supplementary facts to prove the theorem.

Lemma 2 (Okada [6]). *There exists a homogeneous polynomial $H_{k,m}(X,Y)$ of degree $k+m$ over K such that*

$$\frac{1}{X^k} = \sum_{r=0}^{m-1} \binom{k+r-1}{r} \cdot \frac{(Y-X)^r}{Y^{k+r}} + (Y-X)^m H_{k,m}\left(1/X, 1/Y\right).$$

Proof. Since we require knowledge of the construction of $H_{k,m}(X,Y)$ in the proof of the next lemma, let us give a proof.

We prove it by induction on m. When $m=1$, one has

$$\frac{1}{X^k} - \frac{1}{Y^k} = (Y-X)\left(\frac{1}{YX^k} + \frac{1}{Y^2X^{k-1}} + \cdots + \frac{1}{Y^kX}\right).$$

If we put

$$H_{k,1}(X,Y) = XY^k + X^2Y^{k-1} + \cdots + X^kY,$$

then

$$\frac{1}{X^k} = \frac{1}{Y^k} + (Y-X)H_{k,1}\left(1/X, 1/Y\right).$$

We next assume that the claim holds for natural numbers less than $m+1$. Then there exists a homogeneous polynomial $H_{l+1,m}(X,Y)$ of degree $l+m+1$ satisfying

$$\frac{1}{X^{l+1}} = \sum_{r=0}^{m-1} \binom{l+r}{r} \cdot \frac{(Y-X)^r}{Y^{l+r+1}} + (Y-X)^m H_{l+1,m}\left(1/X, 1/Y\right).$$

On both side of the above equation, we multiply by $1/Y^{k-l}$, take sum $\sum_{l=0}^{k-1}$, and multiply by $Y-X$. Then

$$\frac{1}{X^k} - \frac{1}{Y^k} = (Y-X)\sum_{l=0}^{k-1} \frac{1}{Y^{k-l}X^{l+1}}$$

$$= \sum_{l=0}^{k-1}\sum_{r=0}^{m-1} \binom{l+r}{r} \cdot \frac{(Y-X)^{r+1}}{Y^{k+r+1}} + (Y-X)^{m+1}\sum_{l=0}^{k-1} \frac{1}{Y^{k-l}} \cdot H_{l+1,m}\left(1/X, 1/Y\right).$$

We see that

$$H_{k,m+1}(X,Y) := \sum_{l=0}^{k-1} Y^{k-l} H_{l+1,m}(X,Y)$$

is a homogeneous polynomial of degree $k+m+1$. Using $\sum_{l=0}^{k-1}\binom{l+r}{r} = \binom{k+r}{r+1}$, we have

$$\frac{1}{X^k} - \frac{1}{Y^k} = \sum_{r=0}^{m-1} \binom{k+r}{r+1} \cdot \frac{(Y-X)^{r+1}}{Y^{k+r+1}} + (Y-X)^{m+1}H_{k,m+1}\left(1/X, 1/Y\right)$$

$$= \sum_{r'=1}^{m-1} \binom{k+r'-1}{r'} \cdot \frac{(Y-X)^{r'}}{Y^{k+r'}} + (Y-X)^{m+1}H_{k,m+1}\left(1/X, 1/Y\right).$$

$$(r' = r+1)$$

Hence the claim holds for $m+1$. $\square$

Lemma 3

$$s_m(a,c)_\Lambda = \frac{1}{a^m c^m} \sum_{x \in R(c)} \sum_{y \in R(a)} \frac{1}{x^{q-1-m}(x-y)^m} + \frac{s_0(c)_\Lambda}{a^m c^m}.$$

Proof. By Proposition 2, we have

$$s_m(a,c)_\Lambda = \frac{1}{a^m c^m} \sideset{}{'}\sum_{\lambda \in \Lambda} \sum_{\mu \in \Lambda} (\lambda/c)^{-q+1+m} (\lambda/c - \mu/a)^{-m}$$

$$= \frac{1}{a^m c^m} \sum_{x \in R(c)} \sum_{y \in R(a) \cup \{0\}} x^{-q+1+m}(x-y)^{-m}$$

$$= \frac{1}{a^m c^m} \left(\sum_{x \in R(c)} x^{-q+1} + \sum_{x \in R(c)} \sum_{y \in R(a)} x^{-q+1+m}(x-y)^{-m} \right). \qquad \square$$

We are now ready to prove Theorem 1. By Lemma 2, for $x \in R(c)$ and $y \in R(a)$,

$$\frac{1}{x^{q-1-m}} = \sum_{r=0}^{m-1} \binom{q-m+r-2}{r} \cdot \frac{(y-x)^r}{y^{q-1-m+r}} + (y-x)^m H_{q-1-m,m}\left(1/x, 1/y\right).$$

Using $\binom{q-m+r-2}{r} = (-1)^r \binom{m+1}{r}$, we obtain

$$s_m(a,c)_\Lambda - \frac{s_0(c)_\Lambda}{a^m c^m}$$

$$= \frac{1}{a^m c^m} \sum_{x \in R(c)} \sum_{y \in R(a)} \frac{1}{x^{q-1-m}(x-y)^m}$$

$$= \frac{1}{a^m c^m} \sum_{x \in R(c)} \sum_{y \in R(a)} \left\{ \sum_{r=0}^{m-1} (-1)^r \binom{m+1}{r} \cdot \frac{(y-x)^r}{y^{q-1-m+r}} \right.$$

$$\left. + (y-x)^m H_{q-1-m,m}\left(1/x, 1/y\right) \right\} \frac{1}{(x-y)^m}$$

$$= \frac{1}{a^m c^m} \sum_{x \in R(c)} \sum_{y \in R(a)} \left\{ \sum_{r=0}^{m-1} \binom{m+1}{r} \cdot \frac{(-1)^{m-r}}{y^{q-1-(m-r)}(y-x)^{m-r}} \right.$$

$$\left. + (-1)^m H_{q-1-m,m}\left(1/x, 1/y\right) \right\}$$

$$= \sum_{r=0}^{m-1} \binom{m+1}{r} (-1)^{m-r} \left(\frac{s_{m-r}(c,a)_\Lambda}{a^r c^r} - \frac{s_0(a)_\Lambda}{a^m c^m} \right)$$

$$+ \frac{(-1)^m}{a^m c^m} \sum_{x \in R(c)} \sum_{y \in R(a)} H_{q-1-m,m}\left(1/x, 1/y\right) \quad \text{(by Lemma 3)}$$

$$= \sum_{r=0}^{m-1} \frac{(-1)^{m-r} s_{m-r}(c,a)_\Lambda}{a^r c^r} \cdot \binom{m+1}{r} - \frac{s_0(a)_\Lambda}{a^m c^m} \sum_{r=0}^{m-1} \binom{m+1}{r} (-1)^{m-r}$$

$$+ \frac{(-1)^m}{a^m c^m} \sum_{x \in R(c)} \sum_{y \in R(a)} H_{q-1-m,m}\left(1/x, 1/y\right).$$

Here note that $\sum_{x\in R(c)}\sum_{y\in R(a)} H_{q-1-m,m}(1/x,1/y) = 0$. Indeed, by construction of $H_{k,m}(X,Y)$ in Lemma 2, $H_{q-1-m,m}(X,Y)$ is expressed as $H_{q-1-m,m}(X,Y) = \sum_{i=1}^{q-2} a_i X^i Y^{q-1-i}$. By Lemma 1,

$$\sum_{x\in R(c)}\sum_{y\in R(a)} H_{q-1-m,m}(1/x,1/y) = \sum_{i=1}^{q-2} a_i \sum_{y\in R(a)} \frac{1}{y^{q-1-i}} \sum_{x\in R(c)} \frac{1}{x^i} = 0.$$

We also note that $\sum_{r=0}^{m-1}\binom{m+1}{r}(-1)^{m-r} = -m$. Therefore we get

$$s_m(a,c)_\Lambda = \sum_{r=0}^{m-1} \frac{(-1)^{m-r} s_{m-r}(c,a)_\Lambda}{a^r c^r} \cdot \binom{m+1}{r} + \frac{s_0(c)_\Lambda}{a^m c^m} + \frac{m \cdot s_0(a)_\Lambda}{a^m c^m}.$$

Finally we transpose the $r = 0$ term of the right hand side to the left hand side. This completes the proof.

6   Concluding Remarks

So far, we have taken some steps toward building a theory of Dedekind sums for finite fields. It would be intriguing to work on the following questions:

- We can define the Dedekind η function for a finite field (cf. Gekeler [2]). Can we connect it with finite Dedekind sums?
- Can we define higher-dimensional finite Dedekind sums as Zagier did in [10]?

References

1. Apostol, T.M.: Modular Functions and Dirichlet Series in Number Theory. Springer, Heidelberg (1990)
2. Gekeler, E.-U.: Finite modular forms. Finite Fields and Their Applications 7, 553–572 (2001)
3. Goss, D.: The algebraist's upper half-planes. Bull. Amer. Math. Soc. 2, 391–415 (1980)
4. Goss, D.: Basic Structures of Function Fields. Springer, Heidelberg (1996)
5. Hamahata, Y.: Dedekind sums for finite fields. In: Diophantine Analysis and Related Fields: DARF 2007/2008. AIP Conference Proceedings, vol. 976, pp. 96–102 (2008)
6. Okada, S.: Analogies of Dedekind sums in function fields. Mem. Gifu Teach. Coll. 24, 11–16 (1989), http://ci.nii.ac.jp/naid/110004649314/
7. Rademacher, H., Grosswald, E.: Dedekind Sums, The Mathematical Association of America, Washington (1972)
8. Sczech, R.: Dedekindsummen mit elliptischen Funktionen. Invent. Math. 76, 523–551 (1984)
9. Serre, J.-P.: Cours d'arithmétique, Presses Universitaires de France, Paris (1970)
10. Zagier, D.: Higher-dimensional Dedekind sums. Math. Ann. 202, 149–172 (1973)

Transitive q-Ary Functions over Finite Fields or Finite Sets: Counts, Properties and Applications

Marc Mouffron

EADS Secure Networks, France
marc.mouffron@eads.com

Abstract. To implement efficiently and securely good non-linear functions with a very large number of input variables is a challenge. Partially symmetric functions such as transitive functions are investigated to solve this issue. Known results on Boolean symmetric functions are extended both to transitive functions and to q-ary functions (on any set of q elements including finite fields $GF(q)$ for any q). In a special case when the number of variables is $n = p^k$ with p prime, an extension of Lucas' theorem provides new counting results and gives useful properties on the set of transitive functions. Results on balanced transitive q-ary functions are given. Implementation solutions are suggested based on q-ary multiple-valued decision diagrams and examples show simple implementations for these kind of symmetric functions. Applications include ciphers design and hash functions design but also search for improved covering radius of codes.

Keywords: Symmetric functions, (sharply) t-transitive functions, balanced functions, functions over finite fields, hardware and software implementation.

1   Introduction

In information science efficient implementations of functions with a large number of variables combined with some properties is a challenge. For example, the choice of good balanced non-linear functions is essential to ensure the quality of a cryptographic algorithm [16] or to achieve optimality of hash coding function for data storage and retrieval [13]. It is often a puzzle to implement them efficiently.

Why is a large number of variables needed ? In cryptography, CAMION et al. [7] gives the trade-off between the correlation order t, the algebraic degree d of q-ary functions of n variables: $d + t < (q - 1).n$. The Walsh transform enables to show that functions have a minimal bound correlation value decreasing with n. More recent results on algebraic attacks [2,9] require functions with high algebraic immunity. Bounds have also been issued on the algebraic immunity with a given number of variables [2]. MEIER et al. [17] shows that the algebraic immunity of random balanced Boolean functions with n variables is almost always at least equal to $0.22n$. So the larger is n, the more choices will be available

J. von zur Gathen, J.L. Imaña, and Ç.K. Koç (Eds.): WAIFI 2008, LNCS 5130, pp. 19–35, 2008.

and better will be their characteristics: t, d, non-linearity or algebraic immunity. Both restraints and those of analogous properties justify the use of functions with a lot of input variables.

But the technology restricts the possibilities. The biggest single dice memories currently available contain 128 Gbits (flash November 2007). This means that, a Boolean function of not more than 37 input variables, which is rather small, will fit into the largest available physical memory block. A function with a hundred input binary or q-ary variables and even more is often desirable. The problem is developing a technique for the efficient construction of such a q-ary function. Thus we must choose a complex function, which possesses properties that enable a straightforward implementation.

The symmetric and partially symmetric q-ary functions represent really good candidates on this point. Symmetric and transitive functions by their properties show potential good behaviour as their input variables are not distinguishable, this is also another reason for their choice. Some authors recommend the symmetry intrinsically to achieve indistinguishability between all variables. This is an over requirement as transitivity is sufficient from that point of view and allows much more varieties. Examples of transitive functions are rotation symmetric Boolean functions [18,22] or dihedral symmetric Boolean functions [15], that allow for instance to build bent Boolean functions that are not quadratic contrary with bent symmetric Boolean functions that are always quadratic. Lots of authors have studied Boolean symmetric functions [8,9,14,16,21,22,23] or symmetric functions over $GF(p)$ (p prime) [10] and there are actual use in ciphers implementations such as [3] but the generalization to partially symmetric q-ary functions when q>2 provides lots of new possibilities. For implementations issues, there are already some works on the partially symmetric Boolean functions [12] and on the symmetric q-ary functions [5]. These works show that the decision diagrams are well suited to take benefits of the symmetries and we investigated this further on with very good solutions for symmetric q-ary functions.

On the one hand the technological implementation is easy both in software and hardware and on the other hand the properties are relevant. For example some symmetric functions achieve bound on both constraint of algebraic immunity and nonlinearity in the Boolean case [14] and also in the q-ary case [2]. Here we study symmetric and partially symmetric q-ary functions on purpose to use them with a very large number of variables (possibly up to several hundred). Theoretical and experimental results are presented with among others the counting of interesting functions and their constructive design. The applications of t-transitive q-ary functions and symmetric functions are rather wide:

1. in cryptography for both stream ciphers [3] and block ciphers,
2. in hash functions design for cryptography [18] or for hash coding providing optimal data storage and retrieval [13],
3. in coding theory to improve results on covering radius.

Current applications use essentially symmetric Boolean functions, but in all these fields transitive and t-transitive q-ary functions can provide enhancement. This paper is organized as follows. Section 2 presents the definitions. Section 3 introduces the counting results. Section 4 exposes the balance property study with new functions. Section 5 proposes application of our techniques to cryptography. Section 6 concludes this paper.

2 Definitions – Notations

Let E_q (E_m) be a set of q elements (m elements), noted $E_q = \{0, .., q-1\}$. A_n is the set of the alternating permutations of E_n and S_n the set of the permutations of E_n. id is the identity permutation. The symmetry groups G are sub-groups of S_n. $C(n, k)$ is the Binomial coefficient k among n values and the Multinomial coefficient $M(n, r_1, \ldots, r_q)$ means choices of $r_1, \ldots, r_q$ among n values. gcd(n,q) is the greatest common divisor of n and q. $GF(q)$ is the Galois Field of q elements. | S | is the Cardinal of set S. $u(E_q)$ is the cardinal of the group of units in a ring E_q. $(a^n) = (a, .., a)$ is the n-tuple whose all coordinates equal a. wa(x) is the weight of n-tuple x with respect to value a. The characteristic function associated with element w is noted: χ_w.

2.1 Definitions on Group Theory and Functions

Definition 1. *A function f from E_q^n onto E_m is **symmetric** if and only if for every permutation P in S_n: $\forall x = (x_i) \in E_q^n : f((x_{P(i)})) = f((x_i))$*

Definition 2. *A function f from E_q^n onto E_m is **invariant** under a permutation P from S_n if and only if: $\forall x = (x_i) \in E_q^n : f((x_{P(i)})) = f((x_i))$*

The symmetric functions are those invariant under the whole symmetric group S_n. For any function f, consider the set of all permutations under which f is invariant, it is a subgroup, G_f, of S_n, called the *symmetry group* of the function f. An interesting introduction on the symmetry group for Boolean functions was done in [19] and most of their lemma have a wider application than just Boolean functions.

Definition 3. *A function f from E_q^n onto E_m is **partially symmetric** if and only if its symmetry group G_f is not the identity element singleton.*

There are many families of partially symmetric functions, for example:

1. If G_f is isomorphic to S_s, where s <n, then f is said to be an *s-over-n symmetric* function.
2. If G_f is A_n, the Alternating group, then f is said to be an *Alternating* function.
3. If G_f is isomorphic to A_s, where s <n, then f is said to be an *s-over-n Alternating* function.
4. If $G_f = C_n$ is the cyclic group, then f is said to be a *rotation symmetric* function.

5. If $G_f = D_n$ is the dihedral group, then f is said to be a *dihedral symmetric* function.
6. If $G_f = \{\text{id, r}\}$ is the group of the a reversal r, then f is said to be *reversaly symmetric* function.

Definition 4. *[6,11] Let $G \subseteq S(E)$ be a group of permutations on a set E, then:*

1. *The group G is **transitive** on E, if for every $x, y \in E$ there is some $g \in G$ such that $g(x) = y$.*
2. *The group G is **regular**(or **sharply transitive**) on E, if for every $x, y \in E$ there is exactly one $g \in G$ such that $g(x) = y$.*
3. *The group G is **r-regularly transitive** on E, if for every $x, y \in E$ the number of permutations $g \in G$ such that $g(x) = y$ is a constant r.*
4. *The group G is **t-transitive** on E, if G induces a transitive group on the set of all ordered tuples of distinct elements from E.*
5. *The group G is **sharply t-transitive** set E, if G induces a regular group on the set of all ordered tuples of distinct elements from E.*

As consequences we have the following definitions for functions, by transferring the properties of the symmetry group acting on the function's input variables set:

Definition 5. *A function f from E_q^n onto E_m is respectively **transitive, sharply transitive, r-regularly transitive, t-transitive, sharply t-transitive** if and only if the symmetry group of f, G_f is respectively a **transitive, sharply transitive, r-regularly transitive, t-transitive, sharply t-transitive**, group acting on E_n.*

For example the function F1 is a transitive function but not a symmetric function: $F1 : (x_1.x_3) \oplus (x_3.x_2) \oplus (x_2.x_4) \oplus \ldots\ldots\ldots \oplus (x_{n-1}.x_n) \oplus (x_n.x_1)$. Its symmetry group is a cyclic group $<(1324 \ldots \text{n-1n})>$.
The function $F2 : (x_1.x_2) \oplus (x_3.x_4) \oplus (x_1.x_2.x_3) \oplus (x_1.x_2.x_4) \oplus (x_1.x_3.x_4) \oplus (x_2.x_3.x_4)$ is a transitive function not symmetric. Its symmetry group is a non-cyclic transitive group $<(12)(34), (13)(24)>$ on $\{1,2,3,4\}$.

Symmetric functions are interesting as they achieve indistinguishability globally on all the input variables. The transitive functions ensure the indistinguishability between any pair of variables, and the sharply transitive functions are a kind of maximal set among them. The t-transitive functions offer intermediate indistinguishability properties. Another driver for this study is that the results on transitive functions apply to a wide number of different sets of functions.

2.2 Definitions on Partitions

The study of symmetric and partially symmetric functions is closely related to the theory of partitions. ANDREW has given a survey on that theory [1].

Definition 6. *A **partition** $\pi = (\pi_1, \pi_2, \pi_3, ..., \pi_m)$ of the integer n in at most m parts each $\leq b$ is a non-increasing sequence of nonnegative integers,*
$$b \geq \pi_1 \geq \pi_2 \geq \pi_3 \geq \ldots \geq \pi_m \geq 0 \text{ such that: } \sum_{i=1}^{m} \pi_i = N.$$

A *partition* π *can also be represented by the number of repetitions of each value*, $\pi =< 0^{r_0}1^{r_1}2^{r_2}...b^{r_b} >$. *Then:* $\sum_{i=0}^{b} i.r_i = N$ *and* $\sum_{i=0}^{b} r_i = m$.

Let, Part(b,m,N), be the set of partitions of all integers $\leq n$ in at most m parts each $\leq$ b. For any n-tuple x $\in E_q^n$, we associate a unique partition of $\pi(x)$ in Part(q-1,n,n.(q-1)).

3   Counts on Transitive Functions

3.1   Partitions Counting

ANDREW [1] gives the formula to enumerate the partitions. We have:

Lemma 1 ([1]Theorem 3.1). $|Part(b,a,ba)| = C(a+b,a) = C(a+b,b)$.

3.2   Permutations Groups and Orbits

The number of distinct subsets of size t in a set of size n is n.(n-1)...(n-t+1). Thus:

Lemma 2. *A t-transitive group of degree n has an order divisible by:* n.(n-1)...(n-t+1).
   A r-regularly t-transitive group of degree n has order: r.n.(n-1)...(n-t+1).

Lemma 3. *(Combinatorics on the orbits): Consider G a group of degree n that operate on* E_n *and by extension also on* E_q^n *by applying operation on the subscripts. Then for any a, b in* E_q, *any x in* E_q^n :

1. *For any G then:* $|x\,G| \leq |G|$.
2. *For any G then:* $|(a^n)\,G| = 1$.
3. *If G is t-transitive then for j<t and x* $= (a^j b^{n-j})$;
   $xG = xSn$ *so* $|a^j b^{n-j}\,G| = C(n,j)$.
4. *For any G if* $\pi(x) =< 0^{r_q}1^{r_1}2r2...q-1^{r_{q-1}} >$ *then:* $|x\,G| \leq M(n,r1,r2,...,rq)$.

Lemma 4. *If G is a transitive group of degree n then:*

$$\forall a \in E_q \ \forall x \in E_q^n \ \exists W \in \mathbb{N} \ / \ wa(x).|xG| = n.W.$$

Proof. In the spirit of [20] count values different to a in the matrix M containing all the n-tuple of xG, the G orbit of x, first on the rows (left hand side) and secondly on the columns (right hand side).

Each row preserves the a's and so contains the same number of elements different to a, wa(x). The number of rows is $|xG|$. This gives left hand side.

Consider z the first column of M. The number of elements of this first column different to a is, wa(z). Due to transitivity of f, a permutation g from G of the columns of M is equivalent to a permutation of the rows of M. For any i there exists a permutation g from G such that g(i)=1 preserving f. Then: $\sum_{y\in xG} wa(y) = \sum_{y\in xG}(\sum_{i=1}^{n} wa(y_i)) = \sum_{y\in xG}(\sum_{i=1}^{n} wa(y_{g(i)})) = \sum_{i=1}^{n}(\sum_{y\in xG} wa(y_1)) = \sum_{i=1}^{n} wa(z) = n.wa(z)$.

This gives right hand side with W=wa(z).

We can immediately deduce some generalization of Lucas' theorem.

Corollary 1. *If G is a transitive group and $n = p^k$ with p prime then for any $x \neq (a^n)$, we have:*

1. *p divides $|xG|$*
2. *n divides $|xG|$ if there exists b such that $gcd(p, wb(x)) = 1$.*

3.3   The n-i Transitive Functions

The set of symmetric functions is not always a proper subset of the partially symmetric functions sets, due to some Pigeon Hole principle as inputs belong to the finite set E_q.

Proposition 1. *Let suppose $q.(i-1) < n$ then f is an n-i transitive function from E_q^n onto E_m (i.e. G_f the symmetry group of f is n-i transitive) if and only if f is a symmetric function from E_q^n onto E_m.*

Proof. Take $p \in S_n - G_f$. Let $X = (x_1, ..., x_n)$ then as $n > q.(i-1)$ among those n components x_j, one value from E_q is repeated at least i times. Then as G_f is n-i transitive there exist one permutation u in G_f exchanging the remaining n-i x_j with their images by p, $x_{p(j)}$, and for which f is invariant then : p(X) = u(X), and, due to transitivity f(u(X))= f(X). So f(p(X))= f(u(X))= f(X) which means f is invariant under p.

Note: For $t > 5$ the only t-transitive groups are S_n ($n \geq t$) and A_n ($n \geq t + 2$) so high transitivity does not bring new functions.

3.4   The Symmetric Functions

Proposition 2. *The number of symmetric functions from E_q^n onto E_m is: $m^{C(n+q-1,q-1)}$.*

Proof. This results from lemma 1 as each S_n orbits is represented by a partition of Part(q-1,n,n.(q-1)).

3.5   The Alternating Functions

The Alternating group A_n of order $\frac{n!}{2}$ acts transitively on E_q^n [11].

Proposition 3. *The number of Alternating functions from E_q^n onto E_m is:*

$$m^{(C(n+q-1,q-1)+C(q,n))}.$$

Proof. As an alternating function is n-2 transitive, when $q < n$, it is also symmetric and $C(q,n) = 0$. Otherwise i.e. when $n \leq q$, we partition the orbits in 2 classes.

For any $X = (x_1, ..., x_n)$ having at least one value repeated twice, say x_j and x_k. Take $\theta \in S_n - A_n$. Consider the transposition τ exchanging x_j and x_k then

$\theta(X) = \theta \circ \tau(X)$, and, $\sigma = \theta \circ \tau \in A_n$ so $f(\theta(X)) = f(\sigma(X)) = f(X)$. So n-tuple X have orbits $X A_n = X S_n$. There are $C(n + q - 1, q - 1) - C(q, n)$ such orbits. For n-tuple $Y = (y_1, ..., y_n)$ with n different values, there are $C(q, n)$ choices for these n values. Consider any transposition τ, then $S_n = A_n \bigcup \tau A_n$ and $Y S_n = Y A_n \bigcup Y \tau A_n$, $|Y S_n| = n!$, $|A_n| = |\tau A_n| = \frac{n!}{2}$, $|Y A_n| \le |A_n|$ and $|Y \tau A_n| \le |\tau A_n|$ so $|Y A_n| = \frac{n!}{2}$. Such n-tuples Y contribute for $2.C(q, n)$ A_n orbits.

3.6   The (Sharply) t-Transitive Functions

Proposition 4. *Given any sharply transitive permutation group G, the set of G-sharply transitive functions from E_q^n onto E_m contains between $m^{\frac{q^n + q.(n-1)}{n}}$ and $m^{q^n - (q.(q-1).(n-1))}$ functions.*

Proof. If G is sharply transitive then $|G| = n$ and so for any x in E_q^n : $|xG| \le n$. From combinatorics of specials orbits we know in particular that $|a^n\ G| = 1$ and $|a^1 b^{n-1}\ G| = n$, it remains then more than $[(q^n - q - n.q.(q-1))/n]$ other orbits and less than $[q^n - q - n.q.(q - 1)]$. So this gives; a lower bound of the total number of orbits : $q + q.(q - 1) + [(q^n - q - n.q.(q - 1))/n]$ and an upper bound of $q + q.(q - 1) + [(q^n - q - n.q.(q - 1))]$.

Example : $G = \{\rho^i / i = 1\ to\ n\}$ which generates the rotation symmetric q-ary functions which dimension exceed $[q^n + q.(n - 1)]/n$. The exact number is given in 3.7.

These results can be generalized to transitive and t-transitive functions.

Proposition 5. *Given any t-transitive permutation group G of size $|G|$, the set of G t-transitive functions from E_q^n onto E_m contains between:*
$$m^{\frac{q^n + q.(|G| - 1) - q.(q - 1).(\sum_{j=1}^t C(n,j) - t.|G|)}{|G|}} \ and \ m^{q^n - q.(q - 1).(\sum_{j=1}^t C(n,j) - t)} \ functions.$$

Proof. According to lemma 3, for $j < t$ $|a^j b^{n-j} G| = C(n,j)$ and $|xG| \le |G|$ for other x. This infers lower bound when other orbits have size $|G|$, and upper bound when those orbits have size 1.

Proposition 6. *Given n=p with p prime and any sharply transitive permutation group G, the set of G-sharply transitive functions from E_q^n onto E_m contains exactly $m^{\frac{q^n + q.(n - 1)}{n}}$ functions.*

Proof. According to corollary 1, if $x \ne (a^n)$ then $n = p$ divides $|xG|$ and $|xG| \le |G| = n$ so $|xG| = n$. So the number of different G-orbits is then exactly: $q + [(q^n - q)/n]$.

3.7   The Rotation Symmetric Functions

As already mentioned rotation symmetric q-ary function are examples of sharply transitive functions, the cycles index of C_n allows an exact enumeration.

Proposition 7. *The set of rotation symmetric functions from E_q^n onto E_m has cardinal $m^{\Gamma_{q,n}}$ with:*

$$\Gamma_{q,n} = \frac{1}{n}\left(\sum_{d/n} \phi(n/d)\ q^d\right) \text{ where } \phi \text{ is the Euler function}$$

Proof. The cycles index of C_n is taken with the range of value q of E_q.

3.8   The Functions with Dihedral Symmetry Group

The dihedral group D_n of order $2n$ acts transitively on E_q^n, then dihedral symmetric q-ary functions can be enumerated from its properties by using the cycles index of D_n.

Proposition 8. *The set of dihedral symmetric q-ary functions from E_q^n onto E_m has cardinal $m^{\Delta_{q,n}}$ with:*

$$\Delta_{q,n} = \frac{1}{2n}\left(\sum_{d/n} \phi(n/d)\ q^d\right) + \begin{cases} \frac{1}{2}q^k & \text{if } n = 2k-1 \\ \frac{1}{4}(q+1)q^k & \text{if } n = 2k \end{cases}$$

where ϕ is the Euler function.

Proof. The cycles index of D_n is taken with the range of value q of E_q.

4   Balance Property

In this section we consider functions from E_q^n onto E_q (m=q).

4.1   Balanced Symmetric Functions

CONDITION FOR BALANCENESS. The first property to check in a cryptologic algorithm is that the functions are balanced. Then the set of following conditions expresses the balance property on symmetric functions:

$$\forall a \in E_q : \ q^{n-1} = \sum_{\substack{\pi \in Part(q-1,n,n.(q-1)) \text{ and } f_\pi = a}} M(n, r_1, r_2, ..., r_q)$$

EXAMPLE: THE AFFINE FUNCTIONS. In a commutative ring E_q, there are $q.(q-1)$ affine symmetric functions. As the number of units in E_q, is $u(E_q)$ (for example in Z/qZ $u(E_q) = \phi(q)$ and for $GF(p)$, p prime, it is $u(E_q) = p-1$), among them $q.u(E_q)$ are balanced: $L_{a,b}(x) = b.(\sum_{i=0}^{n} x_i) + a$ where $a \in E_q$ and b is a unit of E_q.

These $L_{a,b}$ are the only balanced symmetric 2-ary functions known for every n . The table A in appendix shows that on $GF(q)$ when q=2 for some even n, there is actually no other balanced symmetric function (see [23]).

LOWER BOUNDS. The Boolean case was studied in [23] and the case $m = q = p$ prime in [10], our bound is worse for q prime but applies to any q. The classic properties on binomial and multinomial coefficients supply a lower bound to the number of balanced symmetric functions for m=q whatever q. If q=2. As $C(n,i) = C(n, n - i)$, when n is odd, it is possible to build by a central symmetry around $n/2$, $2(n + 1)/2$ balanced functions. In a more general way, there are similar relations on multinomial coefficients.

Proposition 9. *The number of balanced symmetric functions from E_q^n onto E_q is:*

$$\begin{array}{ll} if \ gcd(n,q) = 1 : & \geq (q!)^{\frac{C(n+q-1,q-1)}{q}} \\ if \ gcd(n,q) \neq 1 : & \geq q.u(E_q) \end{array}$$

Proof. For every value of k, if ρ^k is a circular rotation k over the q-tuple: $M(n, \rho^k(r_1, r_2, ..., r_q)) = M(n, r_1, r_2, ..., r_q)$. When gcd(n,q) = 1, for any such q-tuple having $\sum_{i=1}^{q} r_i = n$, we get q different shifts because the r_i cannot be all equal otherwise the sum means $n = r.q$. The $C(n + q - 1, q - 1)$ partitions of n can be spread among $C(n + q - 1, q - 1)/q$ classes of q elements each. When each class is appointed a permutation of the q elements from E_q a balanced symmetric function is obtained. There are then more than $(q!)^{\frac{C(n+q-1,q-1)}{q}}$ balanced symmetric functions.

Some numerical results are given in appendix.

4.2   t-Transitive Balanced q-Ary Functions

A necessary condition can also be stated for t-transitive balanced functions for special n.

Proposition 10. *If $n = p^k + t - 1$ with p prime and f is a t-transitive balanced function on E_q for every w in E_q then:*

If $gcd(p,q) = 1$: $\sum_{a_1=0}^{q-1} \cdots \sum_{a_{t-1}=0}^{q-1} \left[\sum_{b=0}^{q-1} \chi_w[f((b^{n-t+1}a_1 \ldots a_{t-1}))] \right] = q^{t-1} \bmod p$

If $gcd(p,q) > 1$: $\sum_{a_1=0}^{q-1} \cdots \sum_{a_{t-1}=0}^{q-1} \left[\sum_{b=0}^{q-1} \chi_w[f((b^{n-t+1}a_1 \ldots a_{t-1}))] \right] = 0 \bmod p$

Proof. Consider w in E_q. We take the restrictions of f on t-1 variables:
$^{a_1 a_2 \ldots a_{t-1}} f_{x_n=a_1, x_{n-1}=a_2, \ldots, x_{n-t+1}=a_{t-1}}(y) = f(y \mid a_1 a_2 \ldots a_{t-1})$

The symmetry group of f, G_f, is a t-transitive group. Consider the sub-group H of G_f constituted of the elements from G_f that fix $x_n, x_{n-1}, \ldots x_{n-t+1}$. As G_f is a t-transitive group of degree n then H is transitive of degree $n - t + 1$. Furthermore $^{a_1 a_2 \ldots a_{t-1}} f$ is invariant under H. So applying corollary 1 to H gives p divides $|yH|$. We want to compute :

$$Sum_w = \sum_{x \in E_q^n} \chi_w[f(x)] = \sum_{a_1=0}^{q-1} \cdots \sum_{a_{t-1}=0}^{q-1} \left[\sum_{b=0}^{q-1} {}^{a_1 a_2 \ldots a_{t-1}} Sum_w \right]$$

where : $^{a_1 a_2 \ldots a_{t-1}} Sum_w = \sum_{y \in E_q^{n-t+1}} \chi_w[^{a_1 a_2 \ldots a_{t-1}} f(y)].$

In $a_1 a_2 \ldots a_{t-1} Sum_w$, the sum on y can be grouped by H orbits and applying corollary 1, the only remaining terms are the y with the same value $n-t+1$ times (b^{n-t+1}).

$$a_1 a_2 \ldots a_{t-1} Sum_w \bmod p = \sum_{x \in E_q^{n-t+1}} \chi_w[a_1 a_2 \ldots a_{t-1} f(y)] \bmod p$$

$$= \sum_{b=0}^{q-1} \chi_w[a_1 a_2 \ldots a_{t-1} f(b^{n-t+1}))] \bmod p$$

so

$$Sum_w \bmod p = \sum_{a_1=0}^{q-1} \ldots \ldots \sum_{a_{t-1}=0}^{q-1} \left[\sum_{b=0}^{q-1} \chi_w[a_1 a_2 \ldots a_{t-1} f((b^{n-t+1}))] \right] \bmod p$$

$$Sum_w \bmod p = \sum_{a_1=0}^{q-1} \ldots \ldots \sum_{a_{t-1}=0}^{q-1} \left[\sum_{b=0}^{q-1} \chi_w[f((b^{n-t+1} a_1 .. a_{t-1}))] \right] \bmod p$$

Due to balancedness: $Sum_w = \sum_{x \in E_q^n} \chi_w[f(x)] = q^{n-1}$. As $n-1 = p^k + t - 2$ and p is prime right hand side is $q^{t-1} \bmod p$ when $\gcd(p,q) = 1$ and 0 otherwise.

4.3   Sharply Transitive Balanced Functions

Proposition 11. *Given $n=p$ with p prime and any sharply transitive permutation group G, the set of G-sharply transitive balanced functions from E_q^n onto E_q contains exactly $(q!).M(qr, \overbrace{r, \ldots, r}^{q \text{ times } r})$ functions with $r = \frac{q^{n-1}-1}{n}$.*

Proof. First of all $(f(a^n))_{a \in E_q}$ are a permutation of the set E_q. According to corollary 1, if $x \neq (a^n)$ then p divides $|xG|$ and $|xG| \le |G| = n$ so $|xG| = p$. So the number of G-orbits of size 1 is : $N_0 = q$ and the number of G-orbits of size $p=n$ is : $N_1 = \frac{q^n - q}{n} = q\frac{q^{n-1}-1}{n} = q.r$

Any sharply transitive balanced function permutes the N_0 orbits with $(q!)$ possibilities, and for the N_1 orbits of size p defines uniquely a balanced function from a set of qr elements to a set of q elements. Their total number is then: $(q!).M(qr, r, \ldots, r)$.

4.4   Degree of Transitive Balanced Boolean Functions

When $q = 2$, $n = p^k$ with p prime >2 and f is a transitive balanced function on $GF(2)^n$ then from proposition 10 $(f(0^n), f(1^n))$ is either $(0,1)$ or $(1,0)$. A balanced function has no term of degree n, its maximum degree can only be $n-1$ and we have.

Proposition 12. *If $n = p^k$ with p prime > 2, then exactly half of the G transitive balanced functions from $GF(2)^n$ onto $GF(2)$ are of maximum degree $n-1$ and half of the G transitive balanced functions have algebraic degree strictly less than $n-1$.*

Proof. Due to previous result, either $(f(0^n), f(1^n)) = (0, 1)$ or $(f(0^n), f(1^n)) = (1, 0)$. Due to the transitivity, the coefficient of all the term of degree n-1 are equals either all 0 or all 1. For every i $<n$, there exist $\pi \in G_f$ such that $\pi(i) = n$. So coefficients of degree n-1 are:

$$s_{n-1}^i(f) = \sum_{x \in E_q^n / x_i = 0} f(x) \bmod 2 = \sum_{x \in E_q^n / x_i = 0} f(\pi(x)) \bmod 2$$

$$\text{so} \quad s_{n-1}^i(f) = \sum_{y \in E_q^n / y_n = 0} f(y) \bmod 2 = s_{n-1}^n(f)$$

Then the set of transitive balanced functions from $GF(2)^n$ onto $GF(2)$ can be partitioned in two sets:

Smax = { f transitive balanced functions on E_q with degree $n - 1$ }
Sinf = { f transitive balanced functions on E_q with degree $<n - 1$ }

A bijection can be described between these two sets, we can build a function f_m in Smax from any f_i from Sinf by:

For any $x \neq (a^n)$ then $f_m(x) = f_i(x)$, $f_m(0^n) = f_i(1^n)$ and $f_m(1^n) = f_i(0^n)$

We see that f_m is balanced if and only if f_i is balanced. We have also :

$$s_{n-1}^n(f_m) = \sum_{x \in E_q^n / xn=0} f_m(x) \bmod 2$$

$$s_{n-1}^n(f_m) = \sum_{x \in E_q^n / xn=0} f_i(x) - f_i(0^n) + f_m(0^n) \bmod 2$$

$$s_{n-1}^n(f_m) = s_{n-1}^n(f_i) - f_m(1^n) + f_m(0^n) \bmod 2$$

So $s_{n-1}^n(f_m) = 1$ due to the condition $(f_m(0^n), f_m(1^n)) = (0, 1)$ or $(1, 0)$ and by this definition f_m actually belongs to Smax if and only if f_i belong to Sinf.

The results in table A show that the condition n odd is compulsory but the other condition $n = p^k$ does not seem mandatory. On the other hand, the case $n = 35$ with 262168 balanced functions and only 131072 of degree $n - 1$ shows that n odd is neither sufficient. The proposition 10 can also be used to provide similar results with other special n values for t-transitive functions.

5   Implementation Issues and Solutions

There are two main methods optimising the symmetric or partially symmetric functions implementations:

1. q-ary decision diagrams (rather hardware oriented)
2. orbits counters intermediate (rather software oriented)

5.1   q-Ary Decision Diagrams Implementation

The q-ary decision diagrams (also called multiple-valued decision diagrams) are generalizations of binary decision diagrams [4,12] to represent and implement q-ary functions defined over a commutative ring E_q (see [5]). A QRODD is a q-ary graph canonically associated with any q-ary function as QROBDD to Boolean functions.

Definition 7. *Let* f *be any function from* E_q^n *onto* E_m *in* n *variables. We associate with* f, *its* **profile**, $p(f)$ *a sequence of* $n + 1$ *positive integers and its* **complexity** $s(f)$

$$p(f) = (1, p_1(f), \ldots, p_n(f))$$
$$s(f) = 1 + p_1(f) + \cdots + p_n(f)$$

where $p_i(f)$ *is the number of distinct q-ary functions in* $n - i$ *variables obtained from* f *by substituting all possible q-ary values to the* **first** i *q-ary variables* $x_1, \ldots, x_i$.

This complexity measures the number of different "sub-functions" inside f generated by a sequential affectations of values to the variables which is also the number of vertices of the canonical q-ary graph associated with f.

A general QROBDD verifies $p_i(f) \leq inf(2^i, 2^{2^{n-i}})$ [12], and a general q-ary QRODD of functions from E_q^n onto E_m verifies $p_i(f) \leq inf(q^i, m^{q^{n-i}})$. For symmetric functions f from E_q^n onto E_q, BUTLER [5] gives a loose upper bound of $s(f) \leq C(q+n, q)$ nodes. We study a tighter upperbound for symmetric functions from E_q^n onto E_m.

5.2   Size of ODD of q-Ary Symmetric Functions

Proposition 13. *When* $1 < m < C(q - 1 + n, q - 1), 1 < n$, *the size* $s(f)$ *of the QRODD of a symmetric function* f *from* E_q^n *onto* E_m *is less than:*

$$U(n, m, q) = C(q + t(n, m, q) - 1, q) + \sum_{i=0}^{n-t(n,m,q)} m^{C(q-1+i,q-1)}$$

where $t(n,m,q)$ *is defined in lemma 5 in D.*

Proof. For a symmetric function f from E_q^n onto E_m there are also two inequalities. The first one is more relevant on the upper part of the diagram $p_i(f) \leq C(q - 1 + i, q - 1)$ because the functions only depend on the partitions of $(x_1, \cdots, x_i)$ which can only take $C(q - 1 + i, q - 1)$ different values. On the lower part we have symmetric functions with n-i variables which count gives another bound $p_i(f) \leq m^{C(q-1+n-i,q-1)}$.

For any q, n and $1 < m < C(q-1+n, q-1), 1 < n$ then the upper bound U(n,m,q) on the symmetric functions is:

$$U(n, m, q) = \sum_{i=0}^{t(n,m,q)-1} C(q-1+i, q-1) + \sum_{i=t(n,m,q)}^{n} m^{C(q-1+n-i, q-1)}$$

$$U(n, m, q) = C(q + t(n, m, q) - 1, q) + \sum_{i=0}^{n-t(n,m,q)} m^{C(q-1+i, q-1)}$$

We can note that $U(n, q, q) < C(q + n, q)$, so this bound is strictly better that the bound given by BUTLER [5] for $q = m$.

When the deletion rule is applied, we get ODD with the following result.

Proposition 14. *When $1 < m < C(q - 1 + n, q - 1), 1 < n$ and $t(n,m,q)$ is defined in lemma 5 in D, the size of the ODD of a symmetric function from E_q^n onto E_m is less than:*

$$V(n, m, q) = U(n, m, q) - (n - t(n, m, q) - 1).m$$

The gain is not very important in comparison with QRODD as the symmetric functions are dependent on all the variables. We fail in generalizing this to other sets of transitive functions due to the fact that the restriction of a transitive function of n variables is no more a transitive function of $n - 1$ variables. Alternating functions have the right property, they will be studied in another paper.

5.3 Applications

As an application for instance, if one needs four (resp. two) non-linear balanced functions of 52 (resp. 58) binary variables and seeks among the symmetric Boolean functions to have simple implementation, it is a failure. There is none. However just take a non-linear balanced function among the symmetric 16-ary (resp. 4-ary) functions of 13 (resp. 29) variables from $GF(2^4)^{13}$ onto $GF(2^4)$ (resp. $GF(2^2)^{29}$ onto $GF(2^2)$). Our results (proposition 9) provide at least $1, 34.10^{498752770}$ (resp. $2, 89.10^{1711}$) symmetric non-linear balanced 16-ary (resp. 4-ary) functions of 13 (resp. 29) variables, which replace the four (resp. two) non-linear balanced Boolean functions of 52 (resp. 58) binary variables. These 16-ary (resp. 4-ary) functions can be implemented with a 16-ary (resp. 4-ary) decision diagram with at most 67863915 nodes (resp. 28447 nodes).

6   Conclusion

In this paper known results on Boolean symmetric functions or symmetric functions over $GF(2)$ or $GF(p)$ are extended both to partially symmetric and to q-ary functions on a set of q elements (for any $q \geq 2$). The propositions 9 to 12 show that when $q > 2$ the number of balanced symmetric q-ary functions is much higher than in the Boolean case, and that the choice can be easy to select one with further desirable properties. The good point for results on transitive functions is that they can apply to a wide number of different sets of functions. The results of this paper apply notably to case $q = 2^k$, which is very useful for actual use in cryptographic functions or hash functions design. When the number of input variables is odd, this allows to build lots of non-linear Boolean functions easy to implement.

The result of proposition 12 means, as G the symmetry group is any transitive group, that when the number of input variables is $n = p^k$ with p prime >2, half of symmetric balanced Boolean functions have degree $n-1$, half of rotation symmetric balanced Boolean functions have degree $n-1$, half of dihedral symmetric balanced Boolean functions have degree $n-1$, ...

The sets of t-transitive q-ary functions appear open enough to do trade-off between many relevant properties. It will be interesting to pursue work on implementations and on symmetry groups intermediate between the dihedral group and the alternating group as the set of dihedral symmetric q-ary functions appears to be too large and the set of alternating q-ary functions is too small.

Acknowledgments. Author wishes to thank anonymous referees for their critical comments and their valuable suggestions.

References

1. Andrews, G.E.: The theory of partitions, Encyclopedia of mathematics and its applications, vol. 2. Addison-Wesley Publishing Company, Reading (1976)
2. Ars, G., Faugere, J.-C.: Algebraic Immunities of functions over finite fields, INRIA Rapport de recherche N° 5532 (March 2005)
3. Berbain, C., Billet, O., Canteaut, A., Courtois, N., Debraize, B., Gilbert, H., Goubin, L., Gouget, A., Granboulan, L., Lauradoux, C., Minier, M., Pornin, T., Sibert, H.: DECIM-128, https://www.cosic.esat.kuleuven.be
4. Bryant, R.E.: Graph-Based Algorithms for Boolean Function Manipulation. IEEE Transactions on Computers C35(8), 677–691 (1986)
5. Butler, J.T., Herscovici, D.S., Sasao, T., Barton, R.J.: Average and Worst Case Number of Nodes in Decision Diagrams of Symmetric Multiple-Valued Functions. IEEE Transactions on computers 46(4) (April 1997)
6. Cameron Peter, J.: Permutation Groups. Cambridge Univ. Press, Cambridge (1999)
7. Camion, P., Canteaut, A.: Generalization of Siegenthaler inequality and Schnorr-Vaudenay multipermutations. In: Koblitz, N. (ed.) CRYPTO 1996. LNCS, vol. 1109, pp. 372–386. Springer, Heidelberg (1996)
8. Canteaut, A., Videau, M.: Symmetric Boolean Functions. IEEE Transactions on information theory 51(8), 2791–2811 (2005)

9. Chen, H., Li, J.: Lower Bounds on the Algebraic Immunity of Boolean Functions, http://arxiv.org/abs/cs.CR/0608080
10. Cusick, T., Li, Y., Stanica, P.: Balanced Symmetric Functions over $GF(p)$. IEEE Transactions on information theory 54(3), 1304–1307 (2008)
11. Dixon, J.D., Brian, M.: Permutation Groups. Springer, Heidelberg (1996)
12. Heinrich-Litan, L., Molitor, P.: Least Upper Bounds for the Size of OBDDs Using Symmetry Properties. IEEE Transactions on computers 49(4), 271–281 (2000)
13. Knuth, D.: The art of Computer Programming. Sorting and Searching, vol. 3, pp. 506–542 (1973)
14. Lobanov, M.: Tight bound between nonlinearity and algebraic immunity, Cryptology ePrint Archive, Report 2005/441 (2005), http://eprint.iacr.org/
15. Maitra, S., Sarkar, S., Dalai, D.K.: On Dihedral Group Invariant Boolean Functions. In: Workshop on Boolean Functions Cryptography and Applications, 2007 (BFCA 2007), Paris, France, May 2-3 (2007)
16. Mitchell, C.: Enumerating Boolean functions of cryptographic significance. Journal of cryptology 2(3), 155–170 (1990)
17. Meier, W., Pasalic, E., Carlet, C.: Algebraic attacks and decomposition of Boolean functions. In: Cachin, C., Camenisch, J.L. (eds.) EUROCRYPT 2004. LNCS, vol. 3027, pp. 474–491. Springer, Heidelberg (2004)
18. Pieprzyk, J., Qu, C.X.: Fast Hashing and Rotation-Symmetric Functions. Journal of Universal Computer Science 5(1), 20–31 (1999)
19. Qu, C.X., Seberry, J., Pieprzyk, J.: Relationships between Boolean Functions and symmetry group. In: International Computer Symposium 2000, ISC 2000, pp. 1–7 (2000)
20. Rivest, R., Vuillemin, J.: On recognizing graph properties from adjacency matrices. Theoretical Computer Science 3, 371–384 (1976)
21. Sarkar, P., Maitra, S.: Balancedness and Correlation Immunity of Symmetric Boolean Functions. In: Proc. R.C. Bose Centenary Symposium. Electronic Notes in Discrete Mathematics, vol. 15, pp. 178–183 (2003)
22. Stanica, P., Maitra, S.: Rotation symmetric Boolean Functions: Count and cryptographic properties. In: Proceedings of R.C. Bose Centenary Symposium on Discrete Mathematics and Applications. Indian Statistical Institute, Calcutta (December 2002)
23. Von Zur Gathen, J., Roche, J.R.: Polynomials with two values. Combinatorica 17(3), 345–362 (1997)
24. Yuan, L.: Results on rotation symmetric polynomials over $GF(p)$. Information Sciences 178, 280–286 (2008)

Appendix

A Balanced Functions for m=q=2, n=2 to 63

We tabulate the exact number of balanced symmetric functions from $GF(2)^n$ onto $GF(2)$.

Table 1. Number of Balanced Symmetric or Alternating Boolean Functions

n	All	Degree n-1	Degree n-2	n	All	Degree n-1	Degree n-2
2	2	2	0	33	147456	73728	0
3	4	2	2	34	130	0	0
4	2	0	0	35	262168	131072	0
5	8	4	0	36	2	0	0
6	2	0	0	37	524288	262144	0
7	16	8	0	38	38	32	0
8	6	4	0	39	1048576	524288	0
9	32	16	0	40	2	0	0
10	2	0	0	41	2127872	1063936	0
11	64	32	0	42	2	0	0
12	2	0	0	43	4194304	2097152	0
13	144	72	0	44	134	0	0
14	14	4	4	45	8388608	4194304	0
15	256	128	0	46	2	0	0
16	2	0	0	47	17825792	8912896	524288
17	512	256	0	48	4098	4096	0
18	2	0	0	49	33554432	16777216	0
19	1024	512	0	50	6	0	4
20	6	0	0	51	67108864	33554432	0
21	2048	1024	0	52	2	0	0
22	2	0	0	53	134217728	67108864	0
23	4096	2048	0	54	34	16	16
24	50	0	0	55	268435456	134217728	0
25	8192	4096	0	56	6	4	0
26	6	0	4	57	536870912	268435456	0
27	16384	8192	0	58	2	0	0
28	2	0	0	59	1073741824	536870912	0
29	34816	17408	1024	60	2	0	0
30	2	0	0	61	2415919104	1207959552	0
31	66176	33088	64	62	3998	1996	652
32	6	4	0	63	4294967424	2147483712	64

B   Balanced Functions over GF(3)

We tabulate the number of balanced symmetric functions from $GF(3)^n$ onto $GF(3)$:

Table 2. Number of Balanced Symmetric Functions over GF(3)

n	1	2	3	4	5	6
Lower bound	6	36	6	7776	279936	6
Actual number	6	108	2316	451170	842411124	12616571508

C   Dihedral Symmetric Functions from E_q^n

We tabulate the dimension of dihedral symmetric functions from E_q^n.

Table 3. Orbits of Dihedral Symmetric Functions from E_q^n

q - n	1	2	3	4	5	6	7	8
2	2	3	4	6	8	13	18	30
3	3	6	10	21	39	92	198	498
4	4	10	20	55	136	430	1300	4435
5	5	15	35	120	377	1505	5895	25395
6	6	21	56	231	888	4291	20646	107331
7	7	28	84	406	1855	10528	60028	365260
8	8	36	120	666	3536	23052	151848	1058058
9	9	45	165	1035	6273	46185	344925	2707245
10	10	55	220	1540	10504	86185	719290	6278140
11	11	66	286	2211	16775	151756	1399266	13442286
12	12	78	364	3081	25752	254618	2569788	26942565
13	13	91	455	4186	38233	410137	4496323	51084943
14	14	105	560	5565	55160	638015	7548750	92383305
15	15	120	680	7260	77631	963040	12229560	160386360

D   Lemma for Decision Diagrams

Lemma 5. *For any positive integers q and $n > 1$ if $1 < m < C(q - 1 + n, q - 1), 1 < n$, there is a unique turnpoint $t(n,m,q)$ such that:*

$$C(q - 1 + t(n, m, q) - 1, q - 1) < m^{C(q-1+n-(t(n,m,q)-1),q-1)}$$

$$and \quad C(q - 1 + t(n, m, q), q - 1) \geq m^{C(q-1+n-t(n,m,q),q-1)}$$

Proof. Let note $g(t) = C(q - 1 + t, q - 1)$ and $h(t) = m^{C(q-1+n-t,q-1)}$. It is obvious that g is strictly increasing with t, h is strictly decreasing with t on E_{n+1}. Furthermore, the inequalities $g(1) < h(1)$ and $g(n) > h(n)$ are satisfied because: $g(1) = q < m^q \leq h(1) = m^{C(q-1+n-1,q-1)}$ as $n > 1$, and $h(n) = m < g(n) = C(q - 1 + n, q - 1)$ by hypothesis.

Fast Point Multiplication on Elliptic Curves without Precomputation

Marc Joye

Thomson R&D France
Technology Group, Corporate Research, Security Laboratory
1 avenue de Belle Fontaine, 35576 Cesson-Sévigné Cedex, France
marc.joye@thomson.net

Abstract. Elliptic curves find numerous applications. This paper describes a simple strategy to speed up their arithmetic in right-to-left methods. In certain settings, this leads to a non-negligible performance increase compared to the left-to-right counterparts.

Keywords: Elliptic curve arithmetic, binary right-to-left exponentiation, mixed coordinate systems.

1 Introduction

Elliptic curve point multiplication — namely, the computation of $Q = [k]P$ given a point P on an elliptic curve and a scalar k — is central in almost every non-trivial application of elliptic curves (cryptography, coding theory, computational number theory, . . .). Its efficiency depends on different factors: the field definition, the elliptic curve model, the internal point representation and, of course, the scalar multiplication method itself.

The choice of the field definition impacts the performance of the underlying field arithmetic: addition, multiplication and inversion. There are two types of fields: fields where inversion is relatively fast and fields where it is not. In the latter case, projective coordinates are preferred over affine coordinates to represent points on an elliptic curve. Points can also be represented with their x-coordinate only. Point multiplication is then evaluated via Lucas chains [13]. This avoids the evaluation of the y-coordinate, which may result in improved overall performance.

Yet another technique to speed up the computation is to use additional (dummy) coordinates to represent points [4]. This technique was later refined by considering mixed coordinate systems [6]. The strategy is to add two points where the first point is given in some coordinate system and the second point is given in some other coordinate system, to get the result point in some (possibly different) coordinate system.

Basically, there exist two main families of scalar multiplication methods, depending on the direction scalar k is scanned: left-to-right methods and right-to-left methods [5,10]. Left-to-right methods are often used as they lead to many

J. von zur Gathen, J.L. Imaña, and Ç.K. Koç (Eds.): WAIFI 2008, LNCS 5130, pp. 36–46, 2008.

different generalizations, including windowing methods [8]. In this paper, we are interested in implementations on constrained devices like smart cards. Hence, we restrict our attention to binary methods so as to avoid precomputing and storing (small) multiples of input point P. We evaluate the performance of the classical binary algorithms (left-to-right and right-to-left) in different coordinate systems. Moreover, as the inverse of a point on an elliptic curve can in most cases be obtained for free, we mainly analyze their signed variants [14,15]. Quite surprisingly, we find a number of settings where the right-to-left methods outperform the left-to-right methods. Our strategy is to make use of mixed coordinate systems but, unlike [6], we do this on binary methods for scalar multiplication. Such a strategy only reveals useful for the right-to-left methods because, as will become apparent later, the point addition routine and the point doubling routine may use different input/output coordinate systems. This gives rise to further gains not available for left-to-right methods.

The rest of this paper is organized as follows. In the next section, we introduce some background on elliptic curves and review their arithmetic. We also review the classical binary scalar multiplication methods. In Section 3, we present several known techniques to speed up the point multiplication. In Section 4, we describe fast implementations of right-to-left point multiplication. We analyze and compare their performance with prior methods. Finally, we conclude in Section 5.

2   Elliptic Curve Arithmetic

An *elliptic curve over a field* $\mathbb{K}$ is a plane non-singular cubic curve with a $\mathbb{K}$-rational point [16]. If $\mathbb{K}$ is a field of characteristic $\neq 2, 3$,[1] an elliptic curve over $\mathbb{K}$ can be expressed, up to birational equivalence, by the (affine) Weierstraß equation

$$E_{/\mathbb{K}} : y^2 = x^3 + a_4\, x + a_6 \quad \text{with } \Delta := -(4a_4{}^3 + 27a_6{}^2) \neq 0\,,$$

the rational point being the (unique) point at infinity O. The condition $\Delta \neq 0$ implies that the curve is non-singular.

The set of $\mathbb{K}$-rational points on E is denoted by $E(\mathbb{K})$. It forms a commutative group where O is the neutral element, under the *'chord-and-tangent'* law. The inverse of $P = (x_1, y_1)$ is $-P = (x_1, -y_1)$. The addition of $P = (x_1, y_1)$ and $Q = (x_2, y_2)$ on E with $Q \neq -P$ is given by $R = (x_3, y_3)$ where

$$x_3 = \lambda^2 - x_1 - x_2 \quad \text{and} \quad y_3 = \lambda(x_1 - x_3) - y_1 \tag{1}$$

with

$$\lambda = \begin{cases} \dfrac{y_1 - y_2}{x_1 - x_2} & \text{if } P \neq Q \quad \text{[chord]} \\[2mm] \dfrac{3x_1{}^2 + a_4}{2y_1} & \text{if } P = Q \quad \text{[tangent]} \end{cases}.$$

---

[1] We focus on these fields because inversion can be expensive compared to a multiplication. For elliptic curves over binary fields, a fast point multiplication method without precomputation is available [12].

2.1 Coordinate Systems

To avoid (multiplicative) inversions in the addition law, points on elliptic curves are usually represented with projective coordinate systems.

In *homogeneous coordinates*, a point $P = (x_1, y_1)$ is represented by the triplet $(X_1 : Y_1 : Z_1) = (\theta x_1 : \theta y_1 : \theta)$ for some non-zero $\theta \in \mathbb{K}$, on the elliptic curve $Y^2 Z = X^3 + a_4 X Z^2 + a_6 Z^3$. The neutral element is given by the point at infinity $(0 : \theta : 0)$ with $\theta \neq 0$. Conversely, a projective homogeneous point $(X_1 : Y_1 : Z_1)$ with $Z_1 \neq 0$ corresponds to the affine point $(X_1/Z_1, Y_1/Z_1)$.

In *Jacobian coordinates*, a point $P = (x_1, y_1)$ is represented by the triplet $(X_1 : Y_1 : Z_1) = (\lambda^2 x_1 : \lambda^3 y_1 : \lambda)$ for some non-zero $\lambda \in \mathbb{K}$. The elliptic curve equation becomes

$$Y^2 = X^3 + a_4 X Z^4 + a_6 Z^6 .$$

Putting $Z = 0$, we see that the neutral element is given by $O = (\lambda^2 : \lambda^3 : 0)$. Given the projective Jacobian representation of a point $(X_1 : Y_1 : Z_1)$ with $Z_1 \neq 0$, its affine representation can be recovered as $(x_1, y_1) = (X_1/Z_1^2, Y_1/Z_1^3)$.

2.2 Point Addition

We detail the arithmetic with Jacobian coordinates as they give rise to faster formulæ [9].

Replacing (x_i, y_i) with $(X_i/Z_i^2, Y_i/Z_i^3)$ in Eq. (1) we find after a little algebra that the addition of $P = (X_1 : Y_1 : Z_1)$ and $Q = (X_2 : Y_2 : Z_2)$ with $Q \neq \pm P$ (and $P, Q \neq O$) is given by $R = (X_3 : Y_3 : Z_3)$ where

$$X_3 = R^2 + G - 2V , \quad Y_3 = R(V - X_3) - S_1 G , \quad Z_3 = Z_1 Z_2 H \qquad (2)$$

with $R = S_1 - S_2$, $G = H^3$, $V = U_1 H^2$, $S_1 = Y_1 Z_2^3$, $S_2 = Y_2 Z_1^3$, $H = U_1 - U_2$, $U_1 = X_1 Z_2^2$, and $U_2 = X_2 Z_1^2$ [6]. Let M and S respectively denote the cost of a (field) multiplication and of a (field) squaring. We see that the addition of two (different) points requires $12M + 4S$. When a fast squaring is available, this can also be evaluated with $11M + 5S$ by computing $2Z_1 Z_2 = (Z_1 + Z_2)^2 - Z_1^2 - Z_2^2$ and "rescaling" X_3 and Y_3 accordingly [1].

The doubling of $P = (X_1 : Y_1 : Z_1)$ (i.e., when $Q = P$) is given by $R = (X_3 : Y_3 : Z_3)$ where

$$X_3 = M^2 - 2S , \quad Y_3 = M(S - X_3) - 8T , \quad Z_3 = 2Y_1 Z_1 \qquad (3)$$

with $M = 3X_1^2 + a_4 Z_1^4$, $T = Y_1^4$, and $S = 4X_1 Y_1^2$. Letting c denote the cost of a multiplication by constant a_4, the doubling of a point costs $3M + 6S + 1c$ or $1M + 8S + 1c$ by evaluating $S = 2[(X_1 + Y_1^2)^2 - X_1^2 - T]$ and $Z_3 = (Y_1 + Z_1)^2 - Y_1^2 - Z_1^2$ [1].

Remark that Eq. (3) remains valid for doubling O. We get $[2](\lambda^2 : \lambda^3 : 0) = (\lambda^8 : \lambda^{12} : 0) = O$.

2.3   Point Multiplication

Let $k = \sum_{i=0}^{\ell-1} k_i \, 2^i$ with $k_i \in \{0,1\}$ denote the binary expansion of k. The evaluation of $[k]P$, that is, $P + P + \cdots + P$ (k times) can be carried out as

$$[k]P = \sum_{\substack{0 \le i \le \ell-1 \\ k_i = 1}} [k_i]\left([2^i]P\right) = \sum_{\substack{0 \le i \le \ell-1 \\ k_i = 1}} [2^i]P = \sum_{\substack{0 \le i \le \ell-1 \\ k_i = 1}} P_i \quad \text{with} \begin{cases} P_0 = P \\ P_i = [2]P_{i-1} \end{cases}.$$

By keeping track of the successive values of P_i in a variable R_1 and by using a variable R_0 to store the accumulated value, $\sum P_i$, we so obtain the following right-to-left algorithm:

Algorithm 1. Right-to-left binary method

Input: P, $k \ge 1$
Output: $[k]P$
1: $R_0 \leftarrow O$; $R_1 \leftarrow P$
2: **while** $(k > 1)$ **do**
3:     **if** (k is odd) **then** $R_0 \leftarrow R_0 + R_1$
4:     $k \leftarrow \lfloor k/2 \rfloor$
5:     $R_1 \leftarrow [2]R_1$
6: **end while**
7: $R_0 \leftarrow R_0 + R_1$
8: **return** R_0

There is a similar left-to-right variant. It relies on the obvious observation that $[k]P = [2]\left([k/2]P\right)$ when k is even. Furthermore, since when k is odd, we can write $[k]P = [k']P + P$ with $k' = k - 1$ even, we get:[2]

Algorithm 2. Left-to-right binary method

Input: P, $k \ge 1$, ℓ the binary length k (i.e., $2^{\ell-1} \le k \le 2^\ell - 1$)
Output: $[k]P$
1: $R_0 \leftarrow P$; $R_1 \leftarrow P$; $\ell \leftarrow \ell - 1$
2: **while** $(\ell \ne 0)$ **do**
3:     $R_0 \leftarrow [2]R_0$
4:     $\ell \leftarrow \ell - 1$
5:     **if** $(\mathrm{bit}(k, \ell) \ne 0)$ **then** $R_0 \leftarrow R_0 + R_1$
6: **end while**
7: **return** R_0

---

[2] We denote by $\mathrm{bit}(k, i)$ bit number i of k; bit number 0 being by definition the least significant bit.

3   Boosting the Performance

3.1   Precomputation

The observation the left-to-right binary method relies on readily extends to higher bases. We have:

$$[k]P = \begin{cases} [2^b]\big([k/2^b]P\big) & \text{if } 2^b \mid k \\ [2^b]\big([(k-r)/2^b]P\big) + [r]P \text{ with } r = k \bmod 2^b & \text{otherwise} \end{cases}.$$

The resulting method is called the 2^b-ary method and requires the prior precomputation of $[r]P$ for $2 \le r \le 2^b - 1$. Observe that when r is divisible by a power of two, say $2^s \mid r$, we obviously have $[k]P = [2^s]\big([2^b]\big([(k-r)/2^{b+s}]P\big) + [r/2^s]P\big)$. Consequently, only odd multiples of P need to be precomputed.

Other choices and optimal strategies for the points to be precomputed are discussed in [2,6]. Further generalizations of the left-to-right binary method to higher bases, including sliding-window methods, are comprehensively surveyed in [8].

3.2   Special Cases

As shown in § 2.2, a (general) point addition in Jacobian coordinates costs $11M + 5S$. In the case $Z_2 = 1$, the addition of $(X_1 : Y_1 : Z_1)$ and $(X_2 : Y_2 : 1) = (X_2, Y_2)$ only requires $7M + 4S$ by noting that $Z_2{}^2$, U_1 and S_1 do not need to be evaluated and that $Z_3 = Z_1 H$. The case $Z_2 = 1$ is the case of interest for the *left-to-right* binary method because the same (input) point P is added when $k_i = 1$ (cf. Line 5 in Algorithm 2).

An interesting case for point doubling is when $a_4 = -3$. Intermediate value M (cf. Eq.(3)) can then be computed as $M = 3(X_1 + Z_1{}^2)(X_1 - Z_1{}^2)$. Therefore, using the square-multiply trade-off for computing Z_3, $Z_3 = (Y_1 + Z_1)^2 - Y_1{}^2 - Z_1{}^2$, we see that the cost of point doubling drops to $3M + 5S$. Another (less) interesting case is when a_4 is a small constant (e.g., $a_4 = \pm 1$ or ± 2) in which case $c \approx 0$ and so the point doubling only requires $1M + 8S$.

3.3   Signed-Digit Representation

A well-known strategy to speed up the evaluation of $Q = [k]P$ on an elliptic curves is to consider the *non-adjacent form* (NAF) of scalar k [14]. The NAF is a canonical representation using the set of digits $\{-1, 0, 1\}$ to uniquely represent an integer. It has the property that the product of any two adjacent digits is zero. Among the signed-digit representations with $\{-1, 0, 1\}$, the NAF has the smallest Hamming weight; on average, only one third of its digits are non-zero [15].

When the cost of point inversion is negligible, it is advantageous to input the NAF representation of k, $k = \sum_{i=0}^{\ell} k'_i 2^i$ with $k'_i \in \{-1, 0, 1\}$ and $k'_i \cdot k'_{i+1} = 0$, and to adapt the scalar multiplication method accordingly. For example, in Algorithm 2, Line 5, R_1 is added when $k'_i = 1$ and R_1 is subtracted when $k'_i = -1$. This strategy reduces the average number of point additions in the left-to-right binary method from $(\ell - 1)/2$ to $\ell/3$.

4 Fast Right-to-Left Point Multiplication

In this section, we optimize as much as possible the binary right-to-left method for point multiplication on elliptic curves over fields $\mathbb{K}$ of characteristic $\neq 2, 3$. We assume that inversion in $\mathbb{K}$ is relatively expensive compared to a multiplication in $\mathbb{K}$ and so restrict our attention to inversion-free formulæ.

We do not consider windowing techniques, which require precomputing and storing points. The targets we have in mind are constrained devices. We also wish a general method that works for *all* inputs and elliptic curves. We assume that the input elliptic curve is given by curve parameters a_4 and a_6. We have seen earlier (cf. § 3.2) that the case $a_4 = -3$ is particularly interesting because it yields a faster point doubling. We do not focus on this case because not all elliptic curves over $\mathbb{K}$ can be rescaled to $a_4 = -3$. Likewise, as we consider inversion-free formulæ, we require that the input and output points are given in projective coordinates. This allows the efficient computation of successive point multiplications. In other words, we do not assume *a priori* conditions on the Z-coordinate of input point $\boldsymbol{P}$.

In summary, we are interested in developing of a *fast, compact and general-purpose point multiplication algorithm*.

4.1 Coordinate Systems

In Jacobian coordinates, a (general) point addition requires $11M + 5S$. In [4], Chudnovsky and Chudnovsky suggested to add two more coordinates to the Jacobian representation of points. A point $\boldsymbol{P}$ is given by five coordinates, $(X_1 : Y_1 : Z_1 : E_1 : F_1)$ with $E_1 = Z_1{}^2$ and $F_1 = Z_1{}^3$. This extended representation is referred to as the *Chudnovsky coordinates* and is abbreviated as $\mathcal{J}^c$. The advantage is that the two last coordinates (i.e., E_i and F_i) only need to be computed for the result point, saving $2(S + M) - 1(S + M) = 1M + 1S$ over the classical Jacobian coordinates. In more detail, from Eq. (2), including the square-multiply trade-off and "rescaling", we see that the sum $(X_3 : Y_3 : Z_3 : E_3 : F_3)$ of two (different) points $(X_1 : Y_1 : Z_1 : E_1 : F_1)$ and $(X_2 : Y_2 : Z_2 : E_2 : F_2)$ can now be evaluated as

$$X_3 = R^2 + G - 2V, \quad Y_3 = R(V - X_3) - S_1 G,$$
$$Z_3 = \big((Z_1 + Z_2)^2 - E_1 - E_2\big)H, \quad E_3 = Z_3{}^2, \quad F_3 = E_3 Z_3 \tag{4}$$

with $R = S_1 - S_2$, $G = 4H^3$, $V = 4U_1 H^2$, $S_1 = 2Y_1 F_2$, $S_2 = 2Y_2 F_1$, $H = U_1 - U_2$, $U_1 = X_1 E_2$, and $U_2 = X_2 E_1$, that is, with $10M + 4S$. The drawback of Chudnovsky coordinates is that doubling is slower. It is easy to see from Eq. (3) that point doubling in Chudnovsky coordinates costs one more multiplication, that is, $2M + 8S + 1c$.

A similar approach was taken by Cohen, Miyaji and Ono [6] but to reduce the cost of point doubling (at the expense of a slower point addition). Their idea is to add a fourth coordinate, $W_1 = a_4 Z_1{}^4$, to the Jacobian point representation

$(X_1 : Y_1 : Z_1)$. This representation, called *modified Jacobian representation*, is denoted by $\mathcal{J}^m$. With this representation, on input point $(X_1 : Y_1 : Z_1 : W_1)$, its double, $[2](X_1 : Y_1 : Z_1 : W_1)$, is given by $(X_3 : Y_3 : Z_3 : W_3)$ where the expression of X_3, Y_3 and Z_3 is given by Eq. (3) but where M and W_3 are evaluated using W_1. In more detail, we write

$$X_3 = M^2 - 2S, \quad Y_3 = M(S - X_3) - 8T,$$
$$Z_3 = 2Y_1 Z_1, \quad W_3 = 16TW_1 \tag{5}$$

with $M = 3X_1{}^2 + W_1$, $T = Y_1{}^4$, and $S = 2[(X_1 + Y_1{}^2)^2 - X_1{}^2 - T]$. The main observation is that $W_3 := a_4 Z_3{}^4 = 16Y_1{}^4(a_4 Z_1{}^4) = 16TW_1$. This saves $(2S + 1c) - 1M$. Notice that the square-multiply trade-off cannot be used for evaluating Z_3 since the value of $Z_1{}^2$ is not available. The cost of point doubling is thus $3M + 5S$ whatever the value of parameter a_4. The drawback is that point addition is more costly as the additional coordinate, $W_3 = a_4 Z_3{}^4$, needs to be evaluated. This requires $2S + 1c$ and so the cost of point addition becomes $11M + 7S + 1c$.

The different costs are summarized in Table 1. For completeness, we also include the cost when using affine and projective homogeneous coordinates. For affine coordinates, I stands for the cost of a field inversion.

Table 1. Cost of point addition and doubling for various coordinate systems

System	Point addition	Point doubling	$(a_4 = -3)$
Affine ($\mathcal{A}$)	2M + S + I	2M + 2S + I	—
Homogeneous ($\mathcal{H}$)	12M + 2S	5M + 6S + 1c	7M + 3S
Jacobian ($\mathcal{J}$)	11M + 5S	1M + 8S + 1c	**3M + 5S**
Chudnovsky ($\mathcal{J}^c$)	**10M + 4S**	2M + 8S + 1c	4M + 5S
Modified Jacobian ($\mathcal{J}^m$)	11M + 7S + 1c	**3M + 5S**	—

When using projective coordinates, we see that Chudnovsky coordinates yield the faster point addition and that modified Jacobian coordinates yield the faster point doubling on any elliptic curve. We also see that point doubling in modified Jacobian coordinates is as fast as the fastest $a_4 = -3$ case with (regular) Jacobian coordinates.

4.2   Mixed Representations

Rather than performing the computation in a single coordinate system, it would be interesting to consider mixed representations in the hope to get further gains. This approach was suggested in [6]. For left-to-right windowing methods with windows of width $w \geq 2$, the authors of [6] distinguish three type of operations and consider three coordinate systems $\mathcal{C}^i$, $1 \leq i \leq 3$:

1. intermediate point doubling: $\mathcal{C}^1 \to \mathcal{C}^1, \boldsymbol{R_0} \mapsto [2]\boldsymbol{R_0}$;
2. final point doubling: $\mathcal{C}^1 \to \mathcal{C}^2, \boldsymbol{R_0} \mapsto [2]\boldsymbol{R_0}$;
3. point addition: $\mathcal{C}^2 \times \mathcal{C}^3 \to \mathcal{C}^1, (\boldsymbol{R_0}, \boldsymbol{R_1}) \mapsto \boldsymbol{R_0} + \boldsymbol{R_1}$.

For inversion-free routines (or when the relative speed of I to M is slow), they conclude that the optimal strategy is to choose $\mathcal{C}^1 = \mathcal{J}^m$, $\mathcal{C}^2 = \mathcal{J}$ and $\mathcal{C}^3 = \mathcal{J}^c$.

It is worth remarking that the left-to-right binary method (Algorithm 2) and its different generalizations have in common the use of an accumulator (i.e., $\boldsymbol{R_0}$) that is repeatedly doubled and to which the input point or a multiple thereof is repeatedly added. This explains the choices made in [6]:

- the input representation of the point doubling (i.e., $\mathcal{C}^1$) is the same as the output representation of the point addition routine;
- the output representation of the (final) point doubling routine (i.e., $\mathcal{C}^2$) is the same as the input representation of [the first point of] the point addition routine;
- the input representation of [the second point of] the point addition routine (i.e., $\mathcal{C}^3$) should allow the calculation of output point in representation $\mathcal{C}^1$.

4.3   Right-to-left Methods

Interestingly, the classical right-to-left method (Algorithm 1) is not subject to the same conditions: a same register (i.e., $\boldsymbol{R_1}$) is repeatedly doubled but its value is not affected by the point additions (cf. Line 3). As a result, the doubling routine can use any coordinate system as long as its output gives enough information to enable the subsequent point addition.[3] Formally, letting the three coordinate systems $\mathcal{D}^i$, $1 \le i \le 3$, we require the following conditions on the point addition and the point doubling routines:

1. point addition: $\mathcal{D}^1 \times \mathcal{D}^2 \to \mathcal{D}^1, (\boldsymbol{R_0}, \boldsymbol{R_1}) \mapsto \boldsymbol{R_0} + \boldsymbol{R_1}$;
2. point doubling: $\mathcal{D}^3 \to \mathcal{D}^3, \boldsymbol{R_1} \mapsto [2]\boldsymbol{R_1}$ with $\mathcal{D}^3 \supseteq \mathcal{D}^2$.

The NAF-based approach is usually presented together with the left-to-right binary method. It however similarly applies when scalar k is right-to-left scanned. Indeed, if $k = \sum_{i=0}^{\ell} k_i' 2^i$ denotes the NAF expansion of k, we can write

$$[k]P = \sum_{0 \le i \le \ell} [k_i']([2^i]P) = \sum_{\substack{0 \le i \le \ell \\ k_i' \ne 0}} \mathrm{sgn}(k_i')P_i \quad \text{with} \begin{cases} P_0 = P \\ P_i = [2]P_{i-1} \end{cases} \tag{6}$$

and where $\mathrm{sgn}(k_i')$ denotes the sign of k_i' (i.e., $\mathrm{sgn}(k_i') = 1$ if $k_i' > 0$ and $\mathrm{sgn}(k_i') = -1$ if $k_i' < 0$). Note that our previous analysis on the choice of coordinate systems

---

[3] More generally, we require an efficient conversion from the output representation of the point doubling (say, $\mathcal{D}_3$) and the input representation of [the second point of] the point addition (say, $\mathcal{D}_2$). With the aforementioned (projective) point representations, $\{\mathcal{H}, \mathcal{J}, \mathcal{J}^c, \mathcal{J}^m\}$, for the sake of efficiency, this translates into $\mathcal{D}_3 \supseteq \mathcal{D}_2$, that is, that the coordinate system $\mathcal{D}_2$ is a subset of coordinate system $\mathcal{D}_3$.

on the (regular) right-to-left binary method remains valid for the NAF-based variant.

We are now ready to present our algorithm. The fastest doubling is given by the modified Jacobian coordinates. Hence, we take $\mathcal{D}^3 = \mathcal{J}^m$. It then follows that we can choose $\mathcal{D}^2 = \mathcal{J}^m$ or $\mathcal{J}$. As the latter leads to a faster point addition, we take $\mathcal{D}^2 = \mathcal{J}$. For the same reason, we take $\mathcal{D}^1 = \mathcal{J}$. The inputs of the algorithm are point $\boldsymbol{P} = (X_1 : Y_1 : Z_1)_{\mathcal{J}}$ given in Jacobian coordinates and scalar $k \geq 1$. The output is $[k]\boldsymbol{P} = (X_k : Y_k : Z_k)_{\mathcal{J}}$ also given in Jacobian coordinates. For further efficiency, we use a NAF representation for k and compute it on-the-fly. JacAdd$[(X^*, Y^*, Z^*), (T_1, T_2, T_3)]$ returns the sum of $(X^* : Y^* : Z^*)$ and $(T_1 : T_2 : T_3)$ as per Eq. (2), provided that $(X^* : Y^* : Z^*) \neq \pm(T_1 : T_2 : T_3)$ and $(X^* : Y^* : Z^*), (T_1 : T_2 : T_3) \neq \boldsymbol{O}$. The JacAdd routine should be adapted to address these special cases as is done e.g. in [9, § A.10.5]. ModJacDouble$[(T_1, T_2, T_3, T_4)]$ returns the double of point $(T_1 : T_2 : T_3 : T_4)$ in modified Jacobian coordinates as per Eq. (3).

Algorithm 3. Fast right-to-left binary method

Input: $\boldsymbol{P} = (X_1 : Y_1 : Z_1)_{\mathcal{J}}$, $k \geq 1$
Output: $[k]\boldsymbol{P} = (X_k : Y_k : Z_k)_{\mathcal{J}}$
1: $(X^*, Y^*, Z^*) \leftarrow (1, 1, 0)$; $(T_1, T_2, T_3, T_4) \leftarrow (X_1, Y_1, Z_1, a_4 Z_1{}^4)$
2: **while** $(k > 1)$ **do**
3:     **if** (k is odd) **then**
4:         $u \leftarrow 2 - (k \bmod 4)$; $k \leftarrow k - u$
5:         **if** ($u = 1$) **then**
6:             $(X^*, Y^*, Z^*) \leftarrow$ JacAdd$[(X^*, Y^*, Z^*), (T_1, T_2, T_3)]$
7:         **else**
8:             $(X^*, Y^*, Z^*) \leftarrow$ JacAdd$[(X^*, Y^*, Z^*), (T_1, -T_2, T_3)]$
9:         **end if**
10:    **end if**
11:    $k \leftarrow k/2$
12:    $(T_1, T_2, T_3, T_4) \leftarrow$ ModJacDouble$[(T_1, T_2, T_3, T_4)]$
13: **end while**
14: $(X^*, Y^*, Z^*) \leftarrow$ JacAdd$[(X^*, Y^*, Z^*), (T_1, T_2, T_3)]$
15: **return** (X^*, Y^*, Z^*)

Remember that we are targeting constrained devices (e.g., smart cards). In our analysis, we assume that there is no optimized squaring: $\mathsf{S}/\mathsf{M} = 1$. Also as we suppose general inputs, we also assume $\mathsf{c}/\mathsf{M} = 1$. However, to ease the comparison under other assumptions, we present the cost formulæ in their generality. We neglect field additions, subtractions, tests, etc. as is customary.

As a NAF has on average one third of digits non-zero, the expected cost for evaluating $[k]\boldsymbol{P}$ using Algorithm 3 for an ℓ-bit scalar k is

$$\frac{\ell}{3} \cdot (11\mathsf{M} + 5\mathsf{S}) + \ell \cdot (3\mathsf{M} + 5\mathsf{S}) \approx 13.33\ell\,\mathsf{M} \ . \tag{7}$$

This has to be compared with the $\frac{\ell}{3} \cdot (11M + 5S) + \ell \cdot (1M + 8S + 1c) \approx 15.33\ell\,M$ of the (left-to-right or right-to-left) inversion-free NAF-based binary methods using Jacobian coordinates. We gain *2 field multiplications per bit of scalar k.*

One may argue that Algorithm 3 requires one more temporary (field) variable, T_4. If *two* more temporary (field) variables are available, the classical methods can be sped up by using modified Jacobian representation; in this case, the cost becomes $\frac{\ell}{3} \cdot (11M + 7S + 1c) + \ell \cdot (3M + 5S) \approx 14.33\ell\,M$, which is still larger than $13.33\ell\,M$. If *three* more temporary (field) variables are available, the performance of the *left-to-right* method can be best enhanced by adapting the optimal strategy of [6] as described earlier to the case $w = 1$: Input point $\boldsymbol{P}$ is then represented in Chudnovsky coordinates. This saves $1M + 1S$ in the point addition. As a result, the cost for evaluating $[k]\boldsymbol{P}$ becomes $\frac{\ell}{3} \cdot (10M + 6S + 1c) + \ell \cdot (3M + 5S) \approx 13.67\ell\,M > 13.33\ell\,M$.

Consequently, we see that even when further temporary variables are available, Algorithm 3 outperforms *all* NAF-based inversion-free methods without precomputation. The same conclusion holds true when considering unsigned representations for k. Replacing $\ell/3$ with $(\ell-1)/2$, we obtain $\approx 16\ell\,M$ with the proposed strategy, and respectively $18\ell\,M$, $17.5\ell\,M$ and $16.5\ell\,M$ for the other left-to-right binary methods.

In addition to efficiency, Algorithm 3 presents a couple of further advantages. Like the usual right-to-left algorithm, it is compatible with the NAF computation and does not require the knowledge of the binary length of scalar k ahead of time. Moreover, as doubling is performed using modified Jacobian coordinates, the doubling formula is independent of curve parameter a_4.

For sensitive applications, Algorithm 3 can be protected against SPA-type attacks with almost no penalty using the table-based atomicity technique of [3], as well as against DPA-type attacks using classical countermeasures.[4] Furthermore, because scalar k is right-to-left scanned, Algorithm 3 thwarts the doubling attack described in [7]. Note that, if not properly protected against, all left-to-right point multiplication methods (including the Montgomery ladder) are subject to the doubling attack.

5 Conclusion

This paper presented an optimized implementation for inversion-free point multiplication on elliptic curves. In certain settings, the proposed implementation outperforms all such previously known methods without precomputation. Further, it scans the scalar from the right to left, which offers a couple of additional advantages.

Acknowledgments. I am grateful to the reviewers for useful comments.

---

[4] SPA and DPA respectively stand for "simple power analysis" and "differential power analysis"; see [11].

References

1. Bernstein, D.J., Lange, T.: Explicit-formulas database,
   http://www.hyperelliptic.org/EFD/jacobian.html
2. Bernstein, D.J., Lange, T.: Fast scalar multiplication on elliptic curves. In: Mullen, G., Panario, D., Shparlinski, I. (eds.) 8th International Conference on Finite Fields and Applications, Contemporary Mathematics. American Mathematical Society (to appear)
3. Chevallier-Mames, B., Ciet, M., Joye, M.: Low-cost solutions for preventing simple side-channel analysis: Side-channel atomicity. IEEE Transactions on Computers 53(6), 760–768 (2004)
4. Chudnovsky, D.V., Chudnovsky, G.V.: Sequences of numbers generated by addition in formal groups and new primality and factorization tests. Advances in Applied Mathematics 7(4), 385–434 (1986)
5. Cohen, H.: A Course in Computational Algebraic Number Theory. Graduate Texts in Mathematics, vol. 138. Springer, Heidelberg (1993)
6. Cohen, H., Miyaji, A., Ono, T.: Efficient elliptic curve exponentiation using mixed coordinates. In: Ohta, K., Pei, D. (eds.) ASIACRYPT 1998. LNCS, vol. 1514, pp. 51–65. Springer, Heidelberg (1998)
7. Fouque, P.-A., Valette, F.: The doubling attack - Why upwards is better than downwards. In: Walter, C.D., Koç, Ç.K., Paar, C. (eds.) CHES 2003. LNCS, vol. 2779, pp. 269–280. Springer, Heidelberg (2003)
8. Gordon, D.M.: A survey of fast exponentiation methods. Journal of Algorithms 27(1), 129–146 (1998)
9. IEEE 1363-2000. Standard specifications for public key cryptography. IEEE Standards (August 2000)
10. Knuth, D.E.: The Art of Computer Programming, 2nd edn. Addison-Welsey (1981)
11. Kocher, P., Jaffe, J., Jun, B.: Differential power analysis. In: Wiener, M. (ed.) CRYPTO 1999. LNCS, vol. 1666, pp. 388–397. Springer, Heidelberg (1999)
12. López, J., Dahab, R.: Fast multiplication on elliptic curves over $GF(2^m)$ without precomputation. In: Koç, Ç.K., Paar, C. (eds.) CHES 1999. LNCS, vol. 1717, pp. 316–327. Springer, Heidelberg (1999)
13. Montgomery, P.L.: Speeding the Pollard and elliptic curve methods of factorization. Mathematics of Computation 48(177), 243–264 (1987)
14. Morain, F., Olivos, J.: Speeding up the computations on an elliptic curve using addition-subtraction chains. RAIRO Theoretical Informatics and Applications 24(6), 531–543 (1990)
15. Reitwiesner, G.W.: Binary arithmetic. Advances in Computers 1, 231–308 (1960)
16. Silverman, J.H.: The Arithmetic of Elliptic Curves. Graduate Texts in Mathematics, vol. 106. Springer, Heidelberg (1986)

Optimal Extension Field Inversion in the Frequency Domain

Selçuk Baktır and Berk Sunar

WPI, Cryptography & Information Security Laboratory, Worcester, MA , USA

Abstract. In this paper, we propose an adaptation of the *Itoh-Tsujii algorithm* to the frequency domain for efficient inversion in a class of *Optimal Extension Fields*. To the best of our knowledge, this is the first time a frequency domain finite field inversion algorithm is proposed for elliptic curve cryptography. We believe the proposed algorithm would be well suited especially for efficient low-power hardware implementation of elliptic curve cryptography using affine coordinates in constrained small devices such as smart cards and wireless sensor network nodes.

Keywords: Elliptic curve cryptography, finite fields, inversion, discrete Fourier transform, number theoretic transform.

1 Introduction

An efficient method for computing Montgomery multiplication in the frequency domain, named *discrete Fourier transform (DFT) modular multiplication*, was introduced in [5,6]. With the DFT modular multiplication algorithm, multiplication in $GF(p^m)$ can be achieved with only a linear number of base field $GF(p)$ multiplications in addition to a quadratic number of simpler base field operations such as addition and fixed bitwise rotation for practical values of p and m relevant to elliptic curve cryptography (ECC). Utilizing the DFT modular multiplication algorithm, an efficient and low-area implementation of a frequency domain ECC processor architecture is introduced in [7]. The proposed architecture performs all finite field arithmetic operations in the frequency domain, however avoids inversions through the use of projective coordinates. Even though the DFT modular multiplication algorithm proved efficient for hardware implementation of ECC [7], the memory required for storing the projective point coordinates constitutes a large amount of the circuit area. Projective coordinate representation requires three coordinate values to represent a point, while affine coordinate representation requires only two. This may be a significant drawback for projective coordinate implementations of ECC in tightly constrained devices. Therefore, it is important to have a frequency domain inversion algorithm in order to realize ECC in the affine coordinates potentially yielding lower storage requirement and power consumption. With this work we introduce an adaptation of Itoh-Tsujii inversion [10] to the frequency domain for a class of Optimal Extension Fields (OEF) [1,2] $GF(p^m)$ where the field characteristic is a Mersenne

J. von zur Gathen, J.L. Imaña, and Ç.K. Koç (Eds.): WAIFI 2008, LNCS 5130, pp. 47–61, 2008.

prime $p = 2^n - 1$ or a Mersenne prime divisor $p = (2^n - 1)/t$ for a positive integer t and $m = n$. Our algorithm achieves an extension field inversion with only a single inversion, $O(m \log m)$ multiplications and constant multiplications, $O(m^2 \log m)$ additions and $O(m^2 \log m)$ fixed bitwise rotations in the base field $GF(p)$.

In Section 2, we provide some background information on OEFs and their arithmetic both in the time and frequency domains. In Section 3, we present an adaptation of Itoh-Tsujii inversion for OEFs to the frequency domain which can be used for efficient implementation of ECC in the frequency domain using the affine coordinates.

2   Background

2.1   OEFs and Their Arithmetic

An extension field $GF(p^m)$ is generated by using an mth degree polynomial irreducible over $GF(p)$ and comprises the residue classes modulo the irreducible field generating polynomial. OEFs are a special class of finite extension fields which use a field generating polynomial of the form $f(x) = x^m - w$ and have a *pseudo-Mersenne prime* field characteristic given in the form $p = 2^n \pm c$ with $\log_2 c < \lfloor \frac{n}{2} \rfloor$. The following theorem provides a simple means to identify irreducible binomials that can be used in OEF construction:

Theorem 1. *[13] Let $m \geq 2$ be an integer and $w \in GF(p)^*$. Then the binomial $x^m - w$ is irreducible in $GF(p)[x]$ if and only if the following three conditions are satisfied:*

1. *each prime factor of m divides the order e of w in $GF(p)^*$;*
2. *the prime factors of m do not divide $\frac{p-1}{e}$;*
3. *$p = 1 \bmod 4$ if $m = 0 \bmod 4$.*

In OEFs the pseudo-Mersenne prime field characteristic allows efficient reduction in the base field $GF(p)$ operations and the binary field generating polynomial allows for efficient reduction in the extension field. OEFs are found to be successful in ECC implementations where resources such as computational power and memory are constrained [17]. For representing OEF elements, the standard basis is utilized. An OEF element $A \in GF(p^m)$ is represented in standard basis by a polynomial of degree at most $m - 1$ as follows

$$A = \sum_{i=0}^{m-1} a_i x^i = a_0 + a_1 x + a_2 x^2 + \ldots + a_{m-1} x^{m-1},$$

where $a_i \in GF(p)$ for $0 \leq i \leq m - 1$.

Addition/Subtraction
The addition/subtraction of $A, B \in GF(p^m)$ is performed by adding/subtracting the polynomial coefficients as

$$A \pm B = \sum_{i=0}^{m-1} a_i x^i \pm \sum_{i=0}^{m-1} b_i x^i = \sum_{i=0}^{m-1} (a_i \pm b_i) x^i \ .$$

Multiplication

For $A, B \in GF(p^m)$, the product $C = A \cdot B$ is computed in two steps: the polynomial multiplication

$$C' = A \cdot B = \sum_{i=0}^{2m-2} c'_i x^i \tag{1}$$

and then the modular reduction $C = C' \bmod f(x)$ where the binomial $f(x) = x^m - w$ facilitates efficient reduction.

Inversion

An elegant method for inversion was introduced by Itoh and Tsujii [12]. For $A \in GF(p^m)$, where $A \neq 0$, $B = A^{-1}$ is computed in four steps as follows

1. Compute the exponentiation A^{r-1} in $GF(p^m)$, where $r = \frac{p^m - 1}{p - 1}$;
2. Compute the product $A^r = (A^{r-1}) \cdot A$;
3. Compute the inversion $(A^r)^{-1}$ in $GF(p)$;
4. Compute the product $A^{r-1} \cdot (A^r)^{-1} = A^{-1}$.

For the particular choice of

$$r = \frac{p^m - 1}{p - 1} \ ,$$

A^r belongs to the ground field $GF(p)$ [13]. This allows the inversion in step 3 to be computed in $GF(p)$ instead of the larger field $GF(p^m)$. For the exponentiation A^{r-1} in step 1, the exponent $r - 1$ is expanded as follows

$$r - 1 = \frac{p^m - 1}{p - 1} - 1 = p^{m-1} + p^{m-2} + \ldots + p^2 + p \ .$$

This exponentiation is computed by finding the powers A^{p^i}. The original Itoh-Tsujii algorithm proposes to use a normal basis representation over $GF(2)$ which turns the p^i-th power exponentiations into simple bitwise rotations. In [10] this technique was adapted to work efficiently in the standard basis and it was shown that A^{r-1} can be computed by performing at most $\lfloor \log_2(m-1) \rfloor + HW(m-1) - 1$ multiplications and $\lfloor \log_2(m-1) \rfloor + HW(m-1)$ p^i-th power exponentiations in $GF(p^m)$, where $HW(m)$ denotes the hamming-weight of m. A^{p^i} is the i-th iterate of the *Frobenius map* where a single iterate is defined as $\sigma(A) = A^p$. Using the properties $\sigma(A + B) = \sigma(A) + \sigma(B)$ for any $A, B \in GF(p^m)$ and $\sigma(a) = a^p = a$ for any $a \in GF(p)$, the exponentiation $A^{p^i} = \sigma^i(A)$ can be simplified as

$$A^{p^i} = \left(\sum_{j=0}^{m-1} a_j x^j \right)^{p^i} = \sum_{j=0}^{m-1} (a_j x^j)^{p^i} = \sum_{j=0}^{m-1} a_j x^{jp^i} \ . \tag{2}$$

Theorem 2 shows that A^{p^i} can be computed by a simple scaled permutation of the coefficients in the polynomial representation of A.

Theorem 2. *[3] For an irreducible binomial $f(x) = x^m - w$ defined over $GF(p)$, the following identity holds for an arbitrary positive integer i and $A \in GF(p^m)$,*

$$A^{p^i} = \left(\sum_{j=0}^{m-1} a_j x^j \right)^{p^i} = \sum_{j=0}^{m-1} (a_j \; c_{s_j}) x^{s_j}$$

where $s_j = jp^i \bmod m$ and $c_{s_j} = w^{\frac{jp^i - s_j}{m}}$. Furthermore, the s_j values are distinct for $0 \leq j \leq m - 1$.

Using the method in Theorem 2, exponentiations of degree p^i may be achieved with the help of a lookup table of precomputed c_{s_j} values, using not more than $m - 1$ constant coefficient multiplications. When m is prime, Corollary 1 further simplifies this computation by showing that $s_j = jp^i \bmod m$ in Theorem 2 equals j and hence no permutations occur for the coefficients of A.

Corollary 1. *[3] If $f(x) = x^m - w$ is irreducible over $GF(p)$, m is prime, $x^j \in GF(p)[x]$ and i is an arbitrary positive rational integer, then $(x^j)^{p^i} = w^t x^j$ (mod $f(x)$), where $t = \frac{jp^i - j}{m}$.*

Proof. We need to prove that $jp^i \bmod m = j$, or in other words $m | jp^i - j$. Since $m | (p - 1)$ is a necessary condition for the existence of the irreducible binomial $f(x) = x^m - w$ over $GF(p)$ for a prime m (see the first condition in Theorem 1), m also divides $jp^i - j = j(p^i - 1) = j(p - 1)(p^{i-1} + p^{i-2} + \cdots + p + 1)$. Hence, the proof is complete.   □

2.2 OEF Arithmetic in the Frequency Domain

In this section, we briefly explain previous work on DFT based finite field multiplication for ECC in the frequency domain. For further information, the reader is referred to [5,6,7]. In order to perform OEF arithmetic in the frequency domain, one needs to first represent the operands in the frequency domain. To convert an element in $GF(p^m)$ into its frequency domain representation, the *number theoretical transform* is used.

Number Theoretic Transform
Number theoretic transform (NTT) over a ring, also known as *the DFT over a finite field*, was introduced by Pollard [14]. The NTT computations over $GF(p)$ are defined by utilizing a dth primitive root of unity, denoted by r, from $GF(p)$ or a finite extension of $GF(p)$. For a sequence (a) of length d whose entries are from $GF(p)$, the forward NTT of (a) over $GF(p)$, denoted by (A), can be computed as

$$A_j = \sum_{i=0}^{d-1} a_i r^{ij} \; , \; 0 \leq j \leq d - 1 \; . \tag{3}$$

Here we refer to the elements of (a) and (A) by a_i and A_i, respectively, for $0 \leq i \leq d-1$. Likewise, the inverse NTT of (A) over $GF(p)$ can be computed as

$$a_i = \frac{1}{d} \cdot \sum_{j=0}^{d-1} A_j r^{-ij} \ , \ 0 \leq i \leq d-1 \ . \tag{4}$$

The sequences (a) and (A) are referred to as the *time and frequency domain representations*, respectively, of the same sequence. We would like to caution the reader that for an NTT of length d to exist over $GF(p)$, the condition $d|p-1$ should be satisfied. Note that, in this case, the equality $\mathrm{GCD}(d, p) = 1$ holds for the *greatest common denominator* of d and p, and hence the inverse of d in $GF(p)$, which is needed for the inverse NTT computations, always exists.

Cyclic convolution of two d-element sequences (a) and (b) in the time domain results in another d-element sequence (c) and can be computed as follows:

$$c_i = \sum_{j=0}^{d-1} a_j \, b_{i-j \bmod d} \ , \ \ 0 \leq i \leq d-1 \ . \tag{5}$$

According to the convolution theorem, the above cyclic convolution operation in the time domain is equivalent to the following computation in the frequency domain:

$$C_i = A_i \cdot B_i \ , \ \ 0 \leq i \leq d-1 \ , \tag{6}$$

where (A), (B) and (C) denote the DFTs of (a), (b) and (c), respectively. Hence, cyclic convolution of two d-element sequences in the time domain, with complexity $O(d^2)$, is equivalent to simple pairwise multiplication of the DFTs of these sequences and has a surprisingly low $O(d)$ complexity [8]. Multiplication of two polynomials, as in OEF arithmetic described with (1), is equivalent to the *acyclic (linear) convolution* of the polynomial coefficients. However, if we represent elements of $GF(p^m)$, which are polynomials of degree at most $(m-1)$ with coefficients in $GF(p)$, with at least $d = (2m-1)$ element sequences by appending zeros at the end, then the cyclic convolution of two such sequences will be equivalent to their acyclic convolution and hence give us their polynomial multiplication. Note that, using the convolution property, the polynomial product $c(x) = a(x) \cdot b(x)$ can be computed very efficiently in the frequency domain but the final reduction by the field generating polynomial is not performed. For further multiplications to be performed on the product $c(x)$ in the frequency domain, it needs to be first reduced modulo the field generating polynomial. *DFT modular multiplication* algorithm [5,6], presented with Algorithm 1, performs both polynomial multiplication and modular reduction in the frequency domain and thus makes it possible to perform consecutive modular multiplications in the frequency domain.

An OEF element can be represented as a sequence by taking its ordered coefficients. For instance,

$$a(x) = a_0 + a_1 x + a_2 x^2 + \ldots + a_{m-1} x^{m-1} \ ,$$

which is an element of $GF(p^m)$, can be interpreted as the following $d \geq 2m - 1$ sequence after appending $d - m$ zeros to the right:

$$(a) = (a_0, a_1, a_2, \ldots, a_{m-1}, 0, 0, \ldots, 0) . \tag{7}$$

In this work we are interested in achieving arithmetic operations in the frequency domain for the special class of OEFs $GF(p^m)$ where the field characteristic p is a Mersenne prime divisor $p = (2^n - 1)/t$ for a positive integer t, $m = n$ is a prime number and the irreducible field generating polynomial $f(x) = x^m - 2$ is used. Furthermore, we will use the dth primitive root of unity $r = -2 \in GF(p)$ for the NTT computations which makes the sequence length $d = 2m$, since in this case $r = -2$ is a $(2m)^{th}$ primitive root of unity in $GF((2^n - 1)/t)$. When $p = M_n = 2^n - 1$, multiplication of an n-bit number with integer powers of 2 modulo M_n can be achieved with a simple bitwise left rotation of the n-bit number, e.g. multiplication of an n-bit number with 2^i modulo M_n can be achieved with a simple bitwise left rotation by $i \bmod n$ bits. Similarly, multiplication of an n-bit number with integer powers of -2 modulo M_n can be achieved with a simple bitwise left rotation of the number, in addition to a negation if the power of -2 is odd. Furthermore, negation of an n-bit number z modulo M_n can simply be achieved by flipping all of its n bits, assuming $0 \leq z \leq M_n$. Likewise, when $p = M_n/t = (2^n - 1)/t$ for a positive integer t, all intermediary arithmetic operations can be efficiently achieved using Mersenne number arithmetic modulo M_n and only the final result needs to be reduced modulo M_n/t. Hence, all intermediary multiplications with integer powers of ± 2 can be achieved with a simple bitwise rotation, in addition to a negation if the power of $r = -2$ is odd.

OEF Addition/Subtraction in the Frequency Domain
Due to the *linearity* property of the NTT [8], operations in the time domain such as addition/subtraction and multiplication by a scalar directly map to the frequency domain, i.e., for any two sequences (a) and (b) representing elements of $GF(p^m)$ in the time domain and for any two scalars $y, z \in GF(p)$, $\mathrm{NTT}(y \cdot (a) \pm z \cdot (b)) = y \cdot \mathrm{NTT}((a)) \pm z \cdot \mathrm{NTT}((b))$.

OEF Multiplication in the Frequency Domain
The *DFT modular multiplication* algorithm [5,7,6] (Algorithm 1) performs Montgomery multiplication in $GF(p^m)$ in the frequency domain. To the best of our knowledge, this algorithm is the only algorithm which achieves modular multiplication in the frequency domain for OEFs relevant to ECC. A similar algorithm for integers is presented in a later paper [16] for Montgomery multiplication of large integer operands, e.g. larger than 500 bits in length, to be used in algorithms such as RSA [15]. Since the DFT modular multiplication algorithm runs in the frequency domain, the parameters used in the algorithm are in their frequency domain sequence representations. These parameters are the input operands $a(x), b(x) \in GF(p^m)$, the result $c(x) = a(x) \cdot b(x) \cdot x^{-(m-1)} \in GF(p^m)$, irreducible field generating polynomial $f(x)$, normalized irreducible field generating polynomial $f_N(x) = f(x)/f(0)$, the sequence length d, and the indeterminate

Algorithm 1. DFT modular multiplication algorithm for $GF(p^m)$

Input: $(A) \equiv a(x) \in GF(p^m)$, $(B) \equiv b(x) \in GF(p^m)$
Output: $(C) \equiv a(x) \cdot b(x) \cdot x^{-(m-1)} \in GF(p^m)$

1: **for** $i = 0$ to $d - 1$ **do**
2:     $C_i \leftarrow A_i \cdot B_i$
3: **end for**
4: **for** $j = 0$ to $m - 2$ **do**
5:     $S \leftarrow 0$
6:     **for** $i = 0$ to $d - 1$ **do**
7:         $S \leftarrow S + C_i$
8:     **end for**
9:     $S \leftarrow -S/d$
10:    **for** $i = 0$ to $d - 1$ **do**
11:        $C_i \leftarrow (C_i + F_{N_i} \cdot S) \cdot X_i^{-1}$
12:    **end for**
13: **end for**
14: Return (C)

---

x. The time domain sequence representations of the polynomial parameters are $(a), (b), (c), (f), (f_N)$ and (x), respectively, and their frequency domain sequence representations, i.e. the DFTs of the time domain sequence representations, are $(A), (B), (C), (F), (F_N)$ and (X). For the inputs $a(x) \cdot x^{m-1}$ and $b(x) \cdot x^{m-1}$, both in $GF(p^m)$, the DFT modular multiplication algorithm computes $a(x) \cdot b(x) \cdot x^{m-1} \in GF(p^m)$. Thus, it keeps the Montgomery residue representation intact and allows for further computations in the frequency domain using the same algorithm. For further information on DFT modular multiplication and its hardware implementation for ECC, the reader is referred to [5,6] and [7], respectively.

3   Itoh-Tsujii Inversion in the Frequency Domain

We propose a direct adaptation of the Itoh-Tsujii algorithm to the frequency domain for inversion in OEFs. As described in Section 2.1, Itoh-Tsujii inversion involves a chain of multiplications and Frobenius map computations in $GF(p^m)$ in addition to a single inversion in the base field $GF(p)$. For the required $GF(p^m)$ multiplications we propose using DFT modular multiplication. Since Frobenius map computations can be achieved very easily in the time domain with simple pairwise multiplications, we propose performing the Frobenius map computations in the time domain by applying the inverse NTT. Hence, back and forth conversions are required between the frequency and time domains for the Frobenius map computations.

For efficient computations, we propose using efficient parameters such as the irreducible field generating binomial $f(x) = x^m - 2$, $p = (2^n - 1)/t$ where n is odd and equals the field extension degree m, $d = 2m$, and the dth primitive root of unity as $r = -2$. Theorem 3 proves that for $p = (2^n - 1)$ and $m = n$,

Algorithm 2. Itoh-Tsujii inversion in $GF(p^m)$ in the frequency domain where $p = 2^n - 1$, $n = 13$ and $m = n$ (for $A, B \in GF(p^m)$ and a positive integer i, FrobeniusMap(A, i) denotes $A^{p^i} \in GF(p^m)$ and DFTmul(A, B) denotes the result of the DFT modular multiplication of A and B)

Input: $(A) \equiv a(x) \cdot x^{m-1} \in GF(p^m)$
Output: $(B) \equiv a(x)^{-1} \cdot x^{m-1} \in GF(p^m)$
1: // Compute $M \cdot a(x)^{r-1} \cdot x^{m-1} \in GF(p^m)$
2: $T1 \leftarrow$ FrobeniusMap$(A, 1)$ // $A^{(10)_p}$
3: $T1 \leftarrow$ DFTmul$(T1, A)$ // $A^{(11)_p}$
4: $T2 \leftarrow$ FrobeniusMap$(T1, 2)$ // $A^{(1100)_p}$
5: $T1 \leftarrow$ DFTmul$(T1, T2)$ // $A^{(1111)_p}$
6: $T2 \leftarrow$ FrobeniusMap$(T1, 4)$ // $A^{(11110000)_p}$
7: $T1 \leftarrow$ DFTmul$(T1, T2)$ // $A^{(11111111)_p}$
8: $T2 \leftarrow$ FrobeniusMap$(T1, 4)$ // $A^{(111111110000)_p}$
9: $T1 \leftarrow$ DFTmul$(T2, T1)$ // $A^{(111111111111)_p}$
10: $T2 \leftarrow$ FrobeniusMap$(T1, 1)$ // $A^{(1111111111110)_p}$
11: // Compute $M \cdot a(x)^r \cdot x^{m-1} \in GF(p^m)$
12: $T1 \leftarrow$ DFTmul$(T2, A)$
13: // Compute $M^{-1} \cdot (a(x)^r)^{-1} \in GF(p)$
14: $A^{-r} \leftarrow T1_0^{-1}$
15: // Compute $a(x)^{-1} \cdot x^{m-1} \in GF(p^m)$
16: **for** $i = 0$ to $d - 1$ **do**
17:     $B_i \leftarrow A^{-r} \cdot T2_i$
18: **end for**
19: Return (B)

$f(x) = x^m - 2$ is irreducible over $GF(p)$ for all practical values of p relevant to ECC. Furthermore, in [6] a list of relevant binomials of the form $f(x) = x^m - 2$ are presented and shown to be irreducible over $GF(p)$ for many values of $p = (2^n - 1)/t$.

Theorem 3. [6] For a Mersenne prime $p = 2^n - 1$ and for $m = n$, a binomial of the form $x^m \pm 2^s$, where s is an integer not congruent to 0 modulo n, is irreducible in $GF(p)[x]$ if m is not a Wieferich prime.

As noted in Section 2.2, when $r = -2$ and $p = (2^n - 1)/t$, a modular multiplication in $GF(p)$ with a power of r can be achieved very efficiently with a simple bitwise rotation in addition to a negation if the power is odd. Furthermore, it is shown in [7] that for the case of $r = -2$, odd m and $n = m$, i.e. when the bit length of the field characteristic $p = 2^n - 1$ is equal to the field extension degree, DFT modular multiplication can be optimized by precomputing some intermediary values in the algorithm. Note that when $r = -2$, $p = (2^n - 1)/t$, the field generating polynomial is $f(x) = x^m - 2$ and hence $f_N(x) = -\frac{1}{2} \cdot x^m + 1$, m is odd and $m = n$, the following equalities

$$F_{N_i} = -\frac{1}{2} \cdot (-2)^{mi} + 1 = \begin{cases} -\frac{1}{2} + 1 = \frac{1}{2}, & i \ even \\ \frac{1}{2} + 1, & i \ odd \end{cases} \tag{8}$$

Table 1. List of some parameters for efficient inversion in the frequency domain

n	$p = (2^n - 1)/t$	m	d	r	equivalent binary field size
13	8191/1	13	26	-2	$\sim 2^{169}$
17	131071/1	17	34	-2	$\sim 2^{289}$
19	524287/1	19	38	-2	$\sim 2^{361}$
23	8388607/47	23	46	-2	$\sim 2^{401}$

hold in $GF(p)$ since $(-2)^{mi} \equiv (-2)^{ni} \equiv (-1)^{ni}(2^n)^i \equiv (-1)^{ni}$ (mod p). In this case F_{N_i} has only two distinct values, namely $-\frac{1}{2} + 1 = \frac{1}{2}$ and $\frac{1}{2} + 1$. Hence, $F_{N_i} \cdot S$ in step 11 of Algorithm 1 can attain only two values for any distinct value of S and these values can be precomputed outside the loop avoiding all such computations inside the loop. The precomputations can be achieved efficiently with only one bitwise rotation and one addition. Taking these optimizations into account, in DFT modular multiplication one needs to perform $2m$ multiplications in step 2, $(2m-1)(m-1)$ additions in step 7, $m-1$ constant multiplications in step 9, $m-1$ bitwise rotations and $m-1$ additions for the computations of $F_{N_i} \cdot S$ in step 11, $2m(m-1)$ additions for the additions of C_i with $F_{N_i} \cdot S$ in step 11 and $2m(m-1)$ bitwise rotations for multiplications with X_i^{-1} in step 11, all in $GF(p)$, totaling a complexity of $2m$ multiplications, $m-1$ constant multiplications, $4m^2 - 4m$ additions and $2m^2 - m - 1$ bitwise rotations as presented in Table 2. Remember that, in a d-element NTT over $GF(p)$ with a dth primitive root of unity r, the values of the parameters r, d and p are dependent on each other and the equality $\text{GCD}(d, p) = 1$ holds for the *greatest common denominator* of d and p, and hence the inverse of d in $GF(p)$, required in step 9 of Algorithm 1, always exists. A list of some efficient parameters suited for ECC are given in Table 1.

In Algorithm 2, we present the frequency domain Itoh-Tsujii algorithm exemplarily for the finite field $GF(p^m)$ with $p = 2^{13} - 1$ and $m = 13$. Note in Algorithm 2 that, for $A, B \in GF(p^m)$ and a positive integer i, $FrobeniusMap(A,i)$ denotes the ith Frobenius map of A and equals A^{p^i}, and $DFTmul(A,B)$ denotes the DFT modular multiplication of A and B. A^{r-1} is computed in steps $2 - 10$ of the algorithm with four multiplications and five p^i-th power exponentiations in $GF(p^m)$, by using two temporary variables. However, there is a trade-off between the amount of temporary storage requirement and the required number of multiplications and Frobenius map computations. In the computation of A^{r-1}, one can always minimize the number of required temporary variables to one by using an alternating chain of p-th power exponentiations and multiplications with A, e.g., in Algorithm 2 A^{r-1} can be computed with the following chain of computations $T1 = A^{(10)_p}$, $T1 = A^{(11)_p}$, $T1 = A^{(110)_p}$, $T1 = A^{(111)_p}$, $T1 = A^{(1110)_p}$, $T1 = A^{(1111)_p}$, $T1 = A^{(11110)_p}$, $T1 = A^{(11111)_p}$, $T1 = A^{(111110)_p}$, $T1 = A^{(111111)_p}$, $T1 = A^{(1111110)_p}$, $T1 = A^{(1111111)_p}$, $T1 = A^{(11111110)_p}$, $T1 = A^{(11111111)_p}$, $T1 = A^{(111111110)_p}$, $T1 = A^{(111111111)_p}$, $T1 = A^{(1111111110)_p}$, $T1 = A^{(1111111111)_p}$, $T1 = A^{(11111111110)_p}$, $T1 = A^{(11111111111)_p}$, $T1 = A^{(111111111110)_p}$,

$T1 = A^{(11111111111)_p}$ and $T1 = A^{(1111111111110)_p}$ by performing eleven multiplications with A and twelve p-th power exponentiations in $GF(p^m)$. We would like to note here that DFT modular multiplications in Algorithm 2 keep the Montgomery residue representation intact, but each Frobenius map computation adds an additional factor to the result. However, we will see in detail later in this section that these additional factors cancel out within the algorithm.

Frobenius Map Computations

We have seen in Section 2 that, when the field extension degree m is prime and the field generating polynomial is a binomial, Frobenius map computation in the time domain is a simple fixed pairwise multiplication of the polynomial coefficients. Therefore, in Itoh-Tsujii inversion we will convert a frequency domain sequence to the time domain before computing its Frobenius endomorphism and come back to the frequency domain afterwards as shown in Algorithm 3. For $d = 2m$, since the time domain sequences have zeros as their higher ordered m elements, the NTT computations in Algorithm 3 can be simplified. Furthermore, since $d = 2m$ is composite, the performance of the NTT can be improved by utilizing the fast Fourier transform (FFT) [9] for a single level. We present the equivalent single level FFT computation for the inverse NTT operation with (9), and for the forward NTT computation with (10) and (11). Note that (10) and (11) are equivalent, except for the sign between the two summations. For more information on the FFT in OEFs, the reader is referred to [6,4].

$$a_i = \frac{2}{d} \cdot \sum_{j=0}^{m-1} A_{2j} r^{-2ij} \ , \ 0 \leq i \leq m-1 \ . \tag{9}$$

$$A_j = \sum_{i=0}^{\frac{m-1}{2}} a_{2i} r^{2ij} + r^j \sum_{i=0}^{\frac{m-3}{2}} a_{2i+1} r^{2ij} \ , \ 0 \leq j \leq m-1 \ . \tag{10}$$

$$A_{j+m} = \sum_{i=0}^{\frac{m-1}{2}} a_{2i} r^{2ij} - r^j \sum_{i=0}^{\frac{m-3}{2}} a_{2i+1} r^{2ij} \ , \ 0 \leq j \leq m-1 \ . \tag{11}$$

As mentioned in Corollary 1, when m is prime and $f(x) = x^m - w$ is irreducible over $GF(p)$, the equality $(x^j)^{p^i} = w^t x^j \pmod{f(x)}$, where $t = \frac{jp^i - j}{m}$, holds. Hence, the Frobenius coefficients do not need to be permuted. Furthermore, when $p = (2^n - 1)/t$, $m = n$ is prime and $f(x) = x^m - 2$, the following equality holds for the jth coefficient of the ith iterate of the Frobenius map

$$w^t = 2^{\frac{jp^i - j}{m}} = 2^{\frac{j(p^i-1)}{m}} = 2^{j(p^{i-1}+p^{i-2}+\cdots+p+1)\frac{p-1}{m}} \ .$$

Due to the first condition of Theorem 1, since $f(x) = x^m - 2$ is irreducible in $GF(p)$, $m|\mathrm{ord}(2)$ and hence $m|(p-1)$. Thus, the above Frobenius map coefficients are all powers of 2 and multiplications by these coefficients can be achieved with $m - 1$ simple bitwise rotations as shown in step 5 of Algorithm 3. In Algorithm 3, FrobeniusMapCoefficient(i, j) equals $\frac{j(p^i - 1)}{m}$ mod n and denotes the amount of bitwise left-rotations to be performed on the jth coefficient of the time domain sequence to achieve the ith iterate of the Frobenius map. With all the above mentioned optimizations utilized, the complexity of Algorithm 3 in terms of $GF(p)$ operations is m constant multiplications, $m^2 - 2m + 1$ fixed bitwise rotations and $m^2 - m$ additions for the inverse NTT computation, $m^2 - 2m + 1$ fixed bitwise rotations and m^2 additions/subtractions for the forward NTT computation and $m - 1$ fixed bitwise rotations for the Frobenius map computation, totaling m constant multiplications, $2m^2 - 3m + 1$ fixed bitwise rotations and $2m^2 - m$ additions/subtractions, as given in Table 2. Note that, in Algorithm 2, DFTmul(A, B)

Algorithm 3. Frobenius map computation in $GF(p^m)$ in the frequency domain when $p = (2^n - 1)/t$, and the irreducible field generating polynomial is $f(x) = x^m - 2$ (FrobeniusMapCoefficient$(i, j) = \frac{j(p^i - 1)}{m}$ mod n)

Input: i, $(A) \equiv a(x) \cdot x^{m-1} \in GF(p^m)$
Output: $(B) \equiv (a(x) \cdot x^{m-1})^{p^i} \in GF(p^m)$
1: // Compute the time domain representation (a) of (A) using the inverse NTT
2: $(a) \leftarrow$ InverseNTT$((A))$
3: // Perform pairwise multiplications through simple bitwise rotations
4: **for** $j = 1$ to $m - 1$ **do**
5:     $a_j \leftarrow a_j <<$ FrobeniusMapCoefficient(i, j) // left rotate the bits of a_j
6: **end for**
7: // Compute the frequency domain representation (A) of (a) using the NTT
8: $(A) \leftarrow$ NTT$((a))$
9: Return $((A))$

---

function which computes the DFT Montgomery multiplication of (A) and (B) keeps the Montgomery representation with the multiplicative factor x^{m-1} intact, however FrobeniusMap(A, i) function which computes the ith iterate of the Frobenius endomorphism on (A) adds an additional term to the multiplicative factor x^{m-1}. Remember in Corollary 1 that when m is prime and $f(x) = x^m - 2$, the ith iterate of the Frobenius endomorphism on x^{m-1} results in $(x^{m-1})^{p^i} = 2^t x^{m-1}$ where $t = \frac{(m-1)p^i - (m-1)}{m}$. Through the Frobenius map computations in Algorithm 2, the additional multiplicative factors 2^t accumulate to some value M until the computation of A^{r-1} in step 10. Thus, in step 12, the computed value $T1$ corresponds to some time domain value $M \cdot a(x)^r \cdot x^{m-1}$. Note that the ith coefficient of the NTT of $M \cdot a(x)^r \cdot x^{m-1}$ is equal to $T1_i = M \cdot a(x)^r \cdot r^{i(m-1)}$ and thus $T1_0 = M \cdot a(x)^r$. Hence, $M \cdot a(x)^r \in GF(p)$ can be obtained by looking at the 0th coefficient of $T1$. In step 14, by taking the inverse of $T1_0$, $T1_0^{-1} = M^{-1} \cdot a(x)^{-r}$, rather than the desired value $a(x)^{-r}$, is obtained. However, the M^{-1} factor cancels out in the last step, i.e. in

Table 2. Complexities of Algorithm 1, Algorithm 2, Algorithm 3 and time domain Itoh-Tsujii inversion (ITI) in $GF(p^m)$, when $f(x) = x^m - 2$, $p = (2^n - 1)/t$, $m = n$ is odd and $d = 2m$, in terms of the number of $GF(p)$ operations, where the numbers of required multiplications, constant multiplications, additions/subtractions and rotations in $GF(p)$ are denoted by $\#\mathcal{M}$, $\#\mathcal{CM}$, $\#\mathcal{A}/\mathcal{S}$ and $\#\mathcal{R}$, respectively ($\Delta = \lfloor \log_2(m - 1) \rfloor + HW(m - 1)$).

	$\#\mathcal{M}$	$\#\mathcal{CM}$	$\#\mathcal{A}/\mathcal{S}$	$\#\mathcal{R}$
Algorithm 1:	$2m$	$m - 1$	$4m^2 - 4m$	$2m^2 - m - 1$
Algorithm 2:	$2m\Delta + 4m - 2$	$2m\Delta - \Delta$	$6m^2\Delta - 5m\Delta$	$4m^2\Delta - 4m\Delta$
Algorithm 3:	$-$	m	$2m^2 - m$	$2m^2 - 3m + 1$
ITI (time):	$m^2\Delta - m^2$ $+4m - 2$	$-$	$m^2\Delta - m^2$ $-m\Delta + 2m - 1$	$2m\Delta - m$ $-2\Delta + 2$

step 17, when this false value of A^{-r} corresponding to $M^{-1} \cdot a(x)^{-r}$ is multiplied with the false value of A^{r-1} in $T2$ corresponding to $M \cdot a(x)^{r-1} \cdot x^{m-1}$ to give us the expected correct result which is the frequency domain representation of $a(x)^{-1} \cdot x^{m-1} \in GF(p^m)$.

Inversion in $GF(p)$
We propose using Fermat inversion for performing the single inversion in $GF(p)$ required in step 14 of Algorithm 2. For an n-bit prime p, this inversion can be conducted by taking the $(p - 2)^{nd}$ power of the operand through a square-and-multiply chain with no more than $n - 1$ multiply and $n - 1$ square operations in $GF(p)$.

Complexity of Itoh-Tsujii Inversion in the Frequency Domain
As described with Algorithm 2 for the exemplary finite field $GF(p^m)$ with $p = 2^{13} - 1$ and $m = 13$, Itoh-Tsujii algorithm achieves inversion utilizing a chain of multiplications and Frobenius map computations. We have seen that when the field generating polynomial is $f(x) = x^m - 2$, $p = (2^m - 1)/t$, $d = 2m$ and $r = -2$ is used as the dth primitive root of unity, DFT modular multiplication and Frobenius endomorphism operations can be achieved extremely efficiently with the complexities given in Table 2. Using these complexities and remembering the number of each operation required in Itoh-Tsujii inversion, as given in Section 2, and exemplarily for $GF(p^{13})$ with Algorithm 2, one can obtain the complexity of Itoh-Tsujii inversion in the frequency domain as given in Table 2.

Table 3. Complexities of Itoh-Tsujii inversion in $GF(p^{13})$ in the time and frequency domains in terms of the number of $GF(p)$ operations for $f(x) = x^{13} - 2$ and $p = 2^{13} - 1$

	Frequency Domain	Time Domain
#Multiplications	180	726
#Constant Multiplications	125	–
#Additions/Subtractions	4745	636
#Fixed Rotations	3120	109

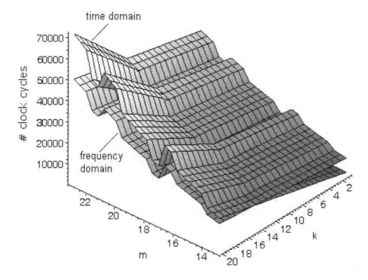

Fig. 1. Number of required clock cycles for inversion in $GF(p^m)$ in the time and frequency domains assuming addition and bitwise-rotation in $GF(p)$ take a single clock cycle and multiplication in $GF(p)$ takes k clock cycles

Multiplication operation is inherently more complex and usually takes more clock cycles to run in hardware. In many modern microprocessors, in order to achieve higher clock rates deeper pipelines are designed in the processor microarchitectures which results in significant differences in the number of clock cycles needed for different instructions. For instance, in the processor microarchitecture of Pentium 4 the latency is only half a clock cycle for a simple 16-bit integer addition, 1 clock cycle for a 32-bit integer addition and 14 clock cycles for a 32-bit integer multiplication [11]. As shown in Table 3 for the exemplary finite field, Itoh-Tsujii algorithm requires a dramatically less number of base field multiplications in the frequency domain than in the time domain. Therefore, it may be desirable to utilize frequency domain inversion in computational environments where multiplication is expensive compared with other operations such as addition and bitwise rotation.

In order to see the cross-over points between the performances of time and frequency domain Itoh-Tsujii algorithms for different multiplication/addition latency ratios k and different field extension degrees m, in Figure 1 we present the total number of clock cycles it takes to achieve inversion with both methods assuming a base field addition/subtraction or bitwise-rotation operation takes only 1 clock cycle to complete and a base field multiplication operation takes k clock cycles. As we can see in the graph, for small multiplication/addition latency ratios inversion in the time domain performs clearly better. For the field extension degree of $m = 13$, the cross-over point is at around $k = 14$, and hence the latency ratio k should be at least 14 for Itoh-Tsujii inversion in $GF(p^{13})$ to perform better in the frequency domain. As the field extension degree m gets larger, frequency domain inversion starts performing better at smaller latency ratios.

4   Conclusion

In this paper we gave an overview of previous work on finite field multiplication in the frequency domain for efficient implementation of ECC. Earlier studies on frequency domain finite field arithmetic lacked inversion, therefore ECC in the frequency domain was implemented only with the projective coordinates with more storage requirement and possibly degraded performance. With this work we proposed an adaptation of the Itoh-Tsujii inversion algorithm to the frequency domain which will make affine coordinate implementation of ECC possible in the frequency domain potentially resulting in less storage requirement and improved performance. To the best of our knowledge, this is the first time a frequency domain inversion algorithm is proposed for implementation of ECC in the frequency domain using affine coordinates.

Acknowledgements

This work was supported by NSF CAREER award ANI-0133297. We would like to thank the anonymous reviewer whose detailed comments and suggestions helped improve the quality of this paper.

References

1. Bailey, D.V., Paar, C.: Optimal Extension Fields for Fast Arithmetic in Public-Key Algorithms. In: Krawczyk, H. (ed.) CRYPTO 1998. LNCS, vol. 1462, pp. 472–485. Springer, Heidelberg (1998)
2. Bailey, D.V., Paar, C.: Efficient Arithmetic in Finite Field Extensions with Application in Elliptic Curve Cryptography. Journal of Cryptology 14(3), 153–176 (2001)
3. Baktır, S.: Efficient Algorithms for Finite Fields, with Applications in Elliptic Curve Cryptography. Master's thesis, Electrical and Computer Engineering Department, Worcester Polytechnic Institute, Worcester, MA, USA (April 2003)

4. Baktır, S., Sunar, B.: Achieving Efficient Polynomial Multiplication in Fermat Fields Using the Fast Fourier Transform. In: Proceedings of the 44th ACM Southeast Conference (ACMSE 2006), March 2006, pp. 549–554. ACM Press, New York (2006)

5. Baktır, S., Sunar, B.: Finite Field Polynomial Multiplication in the Frequency Domain with Application to Elliptic Curve Cryptography. In: Levi, A., Savaş, E., Yenigün, H., Balcısoy, S., Saygın, Y. (eds.) ISCIS 2006. LNCS, vol. 4263, pp. 991–1001. Springer, Heidelberg (2006)

6. Baktır, S., Sunar, B.: Frequency Domain Finite Field Arithmetic for Elliptic Curve Cryptography (preprint, 2007),
http://www.wpi.edu/~selcuk/DFTmultExpanded_preprint.pdf

7. Baktır, S., Kumar, S., Paar, C., Sunar, B.: A State-of-the-art Elliptic Curve Cryptographic Processor Operating in the Frequency Domain. Mobile Networks and Applications (MONET) 12(4), 259–270 (2007)

8. Burrus, C.S., Parks, T.W.: DFT/FFT and Convolution Algorithms. John Wiley & Sons, Chichester (1985)

9. Cooley, J., Tukey, J.: An Algorithm for the Machine Calculation of Complex Fourier Series. Mathematics of Computation 19, 297–301 (1965)

10. Guajardo, J., Paar, C.: Itoh-Tsujii Inversion in Standard Basis and Its Application in Cryptography. Design, Codes, and Cryptography (25), 207–216 (2002)

11. Hinton, G., Sager, D., Upton, M., Boggs, D., Carmean, D., Kyker, A., Roussel, P.: The Microarchitecture of the Pentium 4 Processor. Intel Technology Journal Q1 (2001)

12. Itoh, T., Tsujii, S.: A Fast Algorithm for Computing Multiplicative Inverses in $GF(2^m)$ Using Normal Bases. Information and Computation 78, 171–177 (1988)

13. Lidl, R., Niederreiter, H.: Finite Fields. Encyclopedia of Mathematics and its Applications, vol. 20. Addison-Wesley, Reading (1983)

14. Pollard, J.M.: The Fast Fourier Transform in a Finite Field. Mathematics of Computation 25, 365–374 (1971)

15. Rivest, R.L., Shamir, A., Adleman, L.: A Method for Obtaining Digital Signatures and Public-Key Cryptosystems. Communications of the ACM 21(2), 120–126 (1978)

16. Saldamlı, G., Koç, Ç.K.: Spectral Modular Exponentiation. In: Proceedings of the 18th IEEE Symposium on Computer Arithmetic (2007)

17. Woodbury, A., Bailey, D.V., Paar, C.: Elliptic Curve Cryptography on Smart Cards without Coprocessors. In: IFIP CARDIS 2000, Fourth Smart Card Research and Advanced Application Conference, Bristol, UK, September 20–22, 2000. Kluwer, Dordrecht (2000)

Efficient Finite Fields in the Maxima Computer Algebra System

Fabrizio Caruso, Jacopo D'Aurizio, and Alasdair McAndrew

Dipartimento di Matematica L. Tonelli,
Università di Pisa, Pisa, Italy
caruso@dm.unipi.it
elianto84@gmail.com
http://www.dm.unipi.it

School of Computer Science and Mathematics,
Victoria University, Melbourne, Australia
Alasdair.McAndrew@vu.edu.au
http://sci.vu.edu.au

Abstract. In this paper we present our implementation of finite fields in the free and open Maxima computer algebra system. In the first version of our package we focused our efforts on efficient computation of primitive elements and modular roots. Our optimizations involve some heuristic methods that use "modular composition" and the generalized Tonelli-Shanks algorithm. Other open and free systems such as GP/Pari do not include in their standard packages any support for finite fields. The computation of the primitive element in Maxima is now faster than in Axiom. Our package provides a more user-friendly interface for teaching than other comparable systems.

Keywords: finite fields, primitive element, modular roots, Maxima.

1   Introduction

We present our implementation of efficient arithmetic over finite fields in the free and open computer algebra system Maxima[18]. Our implementation is part of the standard Maxima system since version 5.14. Maxima is a general purpose computer algebra system that is distributed under GPL (GNU General Public License). It is a very old system with a large number of developers and users that contains libraries in many fields of mathematics. It depends internally on a Common Lisp interpreter but at the same time provides a higher level functional language, which we used to implement our library. The first version of the library was based on the paper [17], which describes a package on finite fields for Maxima's ancestor (Macsyma), whose source code is long gone. When used for teaching [12] if compared to Axiom, our package provides a more user-friendly interface.

Our library assumes that a representation of a finite field is given as a vector space over $\mathbb{F}_p$

$$\mathbb{F}_{p^n} := \mathbb{F}_p[x]/\left(f(x)\right),\tag{1}$$

J. von zur Gathen, J.L. Imaña, and Ç.K. Koç (Eds.): WAIFI 2008, LNCS 5130, pp. 62–76, 2008.

where $f(x)$ is a monic and irreducible polynomial of degree n; every element is represented in the monomial (power) base $\left\{x^i\right\}_{i=0}^{n-1}$.

At the moment the library consists of a part dealing with general $\mathbb{F}_{p^n}$ fields and a second part related to modular roots, dealing with $\mathbb{F}_p$ fields.

The main package (gf.mac) contains functions for basic arithmetic operations, exponentiation, computation of the primitive element, discrete logarithm, minimal polynomial, normal bases and linear algebra over finite fields. In order to reduce the time needed to find a primitive element, some techniques that involve "modular composition" ([1] and [21]) and some heuristic methods have been used.

The second part of the package (gf_roots.mac) contains efficient implementations of the Cipolla-Lehmer algorithm based on field extension with the improvements by Müller (see [4], [11], [15], for an equivalent algorithm see [14]) and the Tonelli-Shanks algorithm ([19], [23]). The former is used for the quadratic and cubic case, whereas the latter is used for the cubic and general case.

Most computer algebra systems such as GP/Pari [7] have no native support for finite fields. Other free and open systems such as Axiom [10] do have some support but lack some of the features offered by our package (e.g. modular roots) and the performance in some areas (e.g. primitive element).

2   The Algorithms

The algorithms implemented in the library deal with the main operations related to finite fields, such as basic arithmetic operations, exponentiation, primitive element, logarithms, minimal polynomial, normal bases, linear algebra over finite fields, modular roots. In this section we shortly describe a few of them and give some details on the computation of the smallest primitive element and of modular roots.

2.1   Basic Arithmetic Operations

All the basic operations such as addition, subtraction, inverse, multiplication and division have been implemented. As far as the multiplication of polynomials is concerned, the Karatsuba fast multiplication algorithm (provided by the Maxima built-in command fasttimes) has been used.

2.2   Exponentiation

The exponentiation is done by either repeated squaring or by factoring the exponent and then by repeated squaring on its factors. Factoring the exponent can reduce the number of products required (for exponents whose binary representation has at least four 1s). In practice reducing the number of multiplications is not the only thing to be taken into consideration: squaring and multiplying do not have the same cost and factoring large exponents becomes costly.

Other approaches based on the construction of a suitable addition chain (performing better than the binary one, even by allowing subtractions) do not seem to provide any significant improvement for characteristics of moderate size ($< 2^{31}$).

2.3 Primitive Element

A primitive element is a generator of the multiplicative group $\mathbb{F}^*_{p^n}$, which has order $p^n - 1$. Given a factorization of $p^n - 1 = p_1^{w_1} p_2^{w_2} \dots p_t^{w_t}$, we find the smallest primitive element with respect to our representation by iteratively exponentiating to $(p^n - 1)/p_i$ for $i = 1 \dots t$.

As proven by Wang in [24], if we assume the extended Riemann hypothesis we can find the smallest primitive element with respect to the lexicographic ordering in polynomial time. For more details and more refined bounds we refer to [21].

It is therefore important to optimize the linear case. We do this by "modular composition" in that we follow Brickell's approach [1] i.e. we use some precomputation and Fermat's Little Theorem.

In order to compute $(x + a)^k$ we proceed as follows

1. We (pre-)compute $x^p, x^{2p}, \dots, x^{(n-1)p}$.
2. We (pre-)compute x^{p^i}.
3. We write the exponent k in base p: $k = \sum_{i=0}^m k_i p^i$.
4. We compute $(x + a)^k$:

$$(x + a)^k = (x + a)^{\sum_{i=0}^m k_i p^i} = \prod_{i=0}^m (x^{p^i} + a)^{k_i}. \tag{2}$$

In order to check that an element of $\mathbb{F}^*_{p^n}$ has maximum order through successive non-residuosity tests, when selecting the prime factors $\bar{p}_j$ of $p^n - 1$ that are also factors of $p - 1$ (so that the exponent in base p has all its digits $q_j := (p-1)/\bar{p}_j$), we perform the following operations

$$(x + a)^{\sum_{i=0}^{n-1} q_j p^i} = \left(\prod_{i=0}^{n-1} (x^{p^i} + a) \right)^{q_j} = ((-1)^n f(-a))^{q_j}, \tag{3}$$

saving a great amount of time. When dealing with elements of higher degree (as it probably happens in $\mathbb{F}_{p^n}$, if p is small enough) instead of doing a more costly precomputation we apply the repeated-squaring algorithm.

Heuristics for the Precomputation. Precomputing x^{p^i} has a moderate cost which in a few predictable cases is higher than the improvement that it brings. These few cases occur when the number of distinct primes dividing $p^n - 1$ is smaller than a given threshold, which in our empirical experiments with our implementation is around 3. This is justified by the fact that the precomputation of x^{p^i} is much more effective when the number of tests

$$g^{\frac{p^n - 1}{p_j}} \neq 1$$

required to grant that an element g has maximum order is sufficiently high.

2.4   Discrete Logarithm

The discrete logarithm is computed by using the Pohlig-Hellman algorithm (see [16] for more details). Fundamentally, the algorithm solves the discrete logarithm problem in $\mathbb{F}_{p^n}^*$

$$a^x = b$$

by first factoring

$$p^n - 1 = e_1^{n_1} e_2^{n_2} \cdots e_k^{n_k},$$

then solving the logarithm problem separately for each prime dividing the order of $\mathbb{F}_{p^n}^*$

$$b^y = a^{\frac{p^n-1}{e_i^{n_i}}},$$

and finally collecting all the solutions using the Chinese remainder theorem. The logarithms of the prime powers can be found using a fairly straightforward process. This algorithm is most efficient when all the factors e_i are small. A good account of this algorithm is in Yan [25].

2.5   Minimal Polynomials

Minimal polynomials are computed using a technique described by McEliece [13]. In a field $\mathbb{F}_{p^n}$, the *order* of an element x, denoted $\mathrm{ord}(x)$ is the smallest value m for which $x^m = 1$. The *degree* of x is the smallest positive integer d for which

$$p^d = 1 \pmod{\mathrm{ord}(x)}.$$

The minimal polynomial for x is then

$$f(z) = (z - x)(z - x^p) \cdots (z - x^{p^{d-1}}).$$

2.6   Square Roots in $\mathbb{F}_p$

For the computation of modular square roots we use the Cipolla-Lehmer algorithm ([4], [11], [14]) with the improvements due to Müller [15].

If a is a quadratic residue in $\mathbb{F}_p$ we have $a^{(p-1)/2} \equiv 1 \pmod{p}$, or, denoting with L the Legendre symbol, $L(a, p) = 1$. In such a case, if $L(b^2 - 4a, p) = -1$, the polynomial $x^2 + bx + a$ is irreducible over $\mathbb{F}_p$, with its roots σ and $\bar{\sigma}$ lying in the extension $\mathbb{F}_p[x]/(x^2 + bx + a)$, and satisfying

$$\begin{cases} \sigma + \bar{\sigma} = -b \\ \sigma \bar{\sigma} = a \\ \sigma^p = \bar{\sigma} \end{cases}$$

So it is possible to bring the square root extraction problem into an exponentiation problem over a quadratic extension of the base field, through the following identity:

$$\left(\bar{\sigma}^{\frac{p+1}{2}}\right)^2 = \left(\sigma^{\frac{p+1}{2}}\right)^2 \equiv a \pmod{p}.$$

By defining V_k and M as

$$V_k = \sigma^k + \bar{\sigma}^k; \qquad M = \begin{pmatrix} 0 & -a \\ 1 & -b \end{pmatrix} \tag{4}$$

we have that

$$V_k = \mathrm{Tr}\left(M^k\right). \tag{5}$$

So, by the Cayley-Hamilton theorem

$$V_{n+2} + bV_{n+1} + aV_n = 0,$$

$\{V_i\}$ is a Lucas sequence in $\mathbb{F}_p$ that can be efficiently evaluated in a single point (in our case $V_{\frac{p+1}{2}}$) with a classical repeated-squaring scheme: if $F \in \mathrm{GL}_n(\mathbb{Z})$ is a companion matrix with characteristic polynomial $x^n - \sum_{i=1}^n a_i x^{i-1}$, in the form

$$F e_i = e_{i+1} \text{ for } i \leq n-1,$$

$$F e_n = (a_1, a_2, \ldots, a_n)^T,$$

for $2 \leq m \leq n$ we have

$$(F^k e_m)_1 = a_1 \cdot (F^k e_{m-1})_n,$$

$$(F^k e_m)_i = (F^k e_{m-1})_{i-1} + a_i \cdot (F^k e_{m-1})_n.$$

F is a rank-1 correction of a circulant matrix representing the permutation

$$(x_1, x_2, \ldots, x_{n-1}, x_n) \longrightarrow (x_2, x_3, \ldots, x_n, x_1).$$

Therefore the knowledge of the first column of the matrix F^k allow us to iteratively reconstruct all the elements of F^k, column by column. In the quadratic case, every power of M has the following structure:

$$\begin{pmatrix} A & -aB \\ B & A - bB \end{pmatrix} \tag{6}$$

and there is no need to explicitly store the elements in the second column. In order to perform the computation of M^k (that leads to V_k) through a binary addition chain, we must analyze how the chosen parameters (A, B) transform under the maps

$$\text{(Q-step) } N \longrightarrow N^2$$
$$\text{(QM-step) } N \longrightarrow N^2 \cdot M$$

where N is a structured matrix like in (6). In particular, we start with the identity matrix $(A = 1, B = 0)$, traversing the binary representation of the exponent k from left to right, performing a *Q-step* every time we encounter a bit 0, a *QM-step* every time we encounter a bit 1.

– Q-step:

$$\begin{cases} A \longrightarrow A^2 - aB^2 \\ B \longrightarrow 2AB - bB^2 \end{cases}$$

– QM-step:

$$\begin{cases} A \longrightarrow -aB(2A + bB) \\ B \longrightarrow (A + bB)^2 - aB^2 \end{cases}$$

The number of multiplication (in $\mathbb{F}_p$) needed to perform a single Q-step or QM-step with such a scheme (in our implementation it is known as msqrt) is just 5. However, directly exploiting the property (4) of the Lucas sequence $\{V_i\}$ it is not difficult to argue that

$$\begin{cases} V_{2k} = V_k^2 - 2a^k \\ V_{2k+1} = V_k V_{k+1} + ba^k \end{cases} \tag{7}$$

We may also note that if $p \equiv 3 \pmod 4$ there is no need to work in a field extension, because

$$\left(a^{\frac{p+1}{4}}\right)^2 = a^{\frac{p+1}{2}} = a\, L(a, p) = a.$$

In the difficult case $p \equiv 1 \pmod 4$, having $a = 1$ in (7) would allow to reduce the number of multiplications needed in a single step: this is the main idea of [15] algorithm. Let us define:

$$b_t = at^2 - 2; \qquad p_t(x) = x^2 + b_t x + 1.$$

p_t is irreducible over $\mathbb{F}_p$ iff $L(at^2 - 4, p) = -1$; in that case its roots $\zeta, \bar{\zeta}$ satisfy

$$\zeta(2 - at^2) = \zeta(\zeta + \bar{\zeta}) = \zeta^2 + 1,$$

$$\zeta(4 - at^2) = (\zeta + 1)^2,$$

$$(4 - at^2) \cdot L(at^2 - 4, p) \cdot \zeta^{\frac{p+1}{2}} = (\zeta + 1)(\bar{\zeta} + 1) = 2 + (\zeta + \bar{\zeta}),$$

$$\zeta^{\frac{p+1}{2}} = -1 = \bar{\zeta}^{\frac{p+1}{2}},$$

$$V_{\frac{p-1}{4}}^2 = 2 + \zeta^{\frac{p-1}{2}} + \bar{\zeta}^{\frac{p-1}{2}} = 2 + \bar{\zeta}\zeta^{\frac{p+1}{2}} + \zeta\bar{\zeta}^{\frac{p+1}{2}} = 2 - (\zeta + \bar{\zeta}) = at^2.$$

Hence $t^{-1}V_{\frac{p-1}{4}}$ is a square root of a in $\mathbb{F}_p$ and can be computed in at most $2\log_2(p)$ multiplications in $\mathbb{F}_p$ (an implementation of the Müller algorithm will be included in the next release of our package, available with Maxima 5.15[18]). The hardest part may be to search for a suitable quadratic non-residue in the form $at^2 - 4$; however, under reasonable probabilistic assumptions, the cost of such pre-processing does not affect the general complexity (for details, see [15]).

2.7 Cube Roots in $\mathbb{F}_p$

For the computation of modular cubic roots we have implemented the generalized Tonelli-Shanks and the Cipolla-Lehmer algorithm. The former is preferable in most cases unless a particular sparse irreducible polynomial is known or can be found easily.

Cubic Tonelli-Shanks. For $p \equiv 2 \pmod 3$ the set of cubic residues is the whole $\mathbb{F}_p^*$, and

$$\sqrt[3]{a} = a^{\frac{2p-1}{3}}$$

holds. If $p \equiv 1 \pmod 3$ the set of cubic residues has cardinality $(p-1)/3$, and a is a cubic residue iff $a^{\frac{p-1}{3}} = 1$.[1] If $p \equiv 7 \pmod 9$ we have

$$\left(a^{\frac{p+2}{9}}\right)^3 = a \cdot a^{\frac{p-1}{3}} = a;$$

and for $p \equiv 4 \pmod 9$

$$\left(a^{\frac{2p+1}{9}}\right)^3 = a \cdot \left(a^{\frac{p-1}{3}}\right)^2 = a.$$

In case that $p \equiv 1 \pmod 9$ we consider the partial factorization

$$p - 1 = 3^s q \qquad (s \geq 2, \; 3 \nmid q)$$

and take a cubic non-residue b; b^q has order 3^s in $\mathbb{F}_p^*$. If $q \equiv 1 \pmod 3$ we set $r = a^{\frac{q-1}{3}}$, $t = a r^3$ to have

$$t^{3^s} \equiv 1 \pmod p.$$

As a consequence, there is an index $1 \leq j \leq s$ for which

$$t = (b^q)^{3^j} \quad \text{or} \quad t = (b^q)^{2 \cdot 3^j}.$$

We can find j by counting how many iterations of the map $c(t) : t \to t^3$ are needed to bring the initial value of t to 1; once j is known

$$a = \left(r^{-1} \cdot (b^q)^{3^{j-1}}\right)^3 \quad \text{or} \quad a = \left(r^{-1} \cdot (b^q)^{2 \cdot 3^{j-1}}\right)^3.$$

If $q \equiv -1 \pmod 3$ we set $r = a^{\frac{q+1}{3}}$, $t = a^{-1} r^3$ to obtain, in a similar fashion

$$a = \left(r \cdot (b^q)^{-3^{j-1}}\right)^3 \quad \text{or} \quad a = \left(r \cdot (b^q)^{-2 \cdot 3^{j-1}}\right)^3.$$

This generalization of the original Shanks algorithm, which will be implemented through the function **gf_cbrt** in the next release of our package, is very powerful; we underline that we have already collected a primitive third root of unity in $\mathbb{F}_p^*$, exactly during the non-cubic residue test for b:

$$\omega = \frac{-1 \pm \sqrt{-3}}{2} = (b^q)^{3^{s-1}} = b^{\frac{p-1}{3}}.$$

---

[1] For our purposes we simply perform an exponentiation; an efficient test could be implemented using the generalized Legendre symbol and the cubic reciprocity law: one needs to find two integers q_1, q_2 such that $p = q_1^2 + q_1 q_2 + q_2^2$, for which the function **csplit** is designed.

Cubic Field Extension. In order to have a cube root of a we may also find a monic irreducible polynomial over $\mathbb{F}_p$, $p(x) = x^3 - \sum_{i=1}^{3} a_i x^{i-1}$, with the constant term $a_1 = \pm a$, and compute $V_{\frac{p^2+p+1}{3}}$, or, equivalently, $M^{\frac{p^2+p+1}{3}}$, where M is the companion matrix associated with $p(x)$. However, once we take $p(x)$ in an almost-general form (like $a_3 = 0$), even by choosing a proper representation for the powers of M, we have no chance to achieve a global complexity[2] of $\tilde{O}(k)$ multiplications in $\mathbb{F}_p$. We have developed an algorithm (mcbrt) that computes $V_{\frac{p^2+p+1}{3}}$ through repeated squarings, under the assumption[3] $p(x) = x^3 + bx + a$, operating only on $A = (M^k)_{(2,2)}, B = (M^k)_{(2,3)}, C = (M^k)_{(3,2)}$ as follows:

- Q-step:

$$\begin{cases} A \longrightarrow A^2 + C(2B + bC) \\ B \longrightarrow 2AB - aC^2 \\ C \longrightarrow 2AC + a^{-1}(B + bC)^2 \end{cases}$$

- QM-step:

$$\begin{cases} A \longrightarrow 2AB - aC^2 \\ B \longrightarrow (B + bC)^2 - 2aAC - b(A^2 + C(2B + bC)) \\ C \longrightarrow A^2 + C(2B + bC) \end{cases}$$

The weight of a single step (both Q and QM) is 10 multiplications in $\mathbb{F}_p$. In order to develop something more efficient (magnitude $\tilde{O}(k)$ multiplications) one should look for *small* irreducible polynomials (trinomials, or polynomials with the most of their coefficients lying in the interval $[-2, 2]$): to search for such elements could be quite expensive, while the generalized Shanks algorithm works at comparable speed, requiring only a cubic non-residue to be given. Unsurprisingly enough, in the quadratic case we notice an almost opposite behavior, with field extension approach having shorter running times than the original Tonelli-Shanks algorithm; we may also conjecture that the quadratic case is the only for which field-extension, matrix-based methods are really useful.

2.8   Generalized Shanks Algorithm for k-th Roots

Assuming that k is an odd prime, we have that the k-th root extraction problem in $\mathbb{F}_p$ has a certain number of trivial cases:

- If $p \not\equiv 1 \pmod{k}$, every element of $\mathbb{F}_p^*$ is a k-th residue; by taking h as the inverse of k in $\mathbb{Z}_{(p-1)}^*$ we have $(a^h)^k = a$.
- If $p \equiv 1 \pmod{k}$ but $p \not\equiv 1 \pmod{k^2}$, a is a k-th residue iff $a^{\frac{p-1}{k}} \equiv 1 \pmod{p}$; by taking h as the inverse of $\frac{1-p}{k}$ in $\mathbb{F}_p^*$ we have

$$\left(a^{\frac{hp+(k-h)}{k^2}} \right)^k = a \cdot \left(a^{\frac{p-1}{k}} \right)^h = a.$$

---

[2] We recall that the $\tilde{O}$-notation hides factors that are logarithmic in the argument, for example $3k \log(k) \log(\log(k)) \in \tilde{O}(k)$.

[3] We recall that an efficient irreducibility test for cubic polynomials over $\mathbb{F}_p$ is the Stickelberger criterion [22]: $p(x) = x^3 + bx + a$ is irreducible iff its discriminant D is a quadratic residue and $2(\sqrt{D} + 3a\,\omega)^2$ is not a cubic residue.

In the last case, $p \equiv 1 \pmod{k^2}$, if a is a k-th residue we find s and q such that

$$p - 1 = k^s q, \qquad s \geq 2, \ k \nmid q,$$

then look for a prime b that is not a k-th residue (it is useless to test composite integers: if all the integer divisors of b are k-th residues, b is a k-th residue too) to compute b^q, a primitive k^s-th root of unity. By denoting with $\bar{q}$ the smallest integer for which $q \equiv \bar{q} \pmod{p}$, and defining r and t as

$$r = a^{\frac{q - \bar{q}}{k}}, \qquad t = a^{\bar{q}} \cdot r^k,$$

we have that the repeated application of the map $\varphi_k(x) : x \to x^k$ leads to the determination of two integers i and w satisfying

$$a^{\bar{q}} \cdot r^k = t = (b^q)^{w \cdot k^i}, \qquad 1 \leq i \leq s, \quad w < k.$$

So the problem is solved up to an inversion and a $\bar{q}$-th root extraction in $\mathbb{F}_p$; denoting with $C(k)$ the number of multiplications needed to find a k-th root we have

$$C(k) \leq k + \log_2(p) + C(k/2) \leq 2(k + \log_2(k) \log_2(p)).$$

An implementation of the generalized Shanks algorithm will appear in the next release of our Maxima package.

3   The Library

In order to use the library it is enough to use Maxima 5.14 or any above version. Some routines related to general modular roots, which we have discussed in the previous section, will be available together with other features and improvements only starting from Maxima 5.15.

The library is written in the Maxima high level language. Its source code is distributed together with Maxima and it is found in

`<maxima_path>/share/contrib/gf/`.

Here we give a short description of only some of the functions; for a more detailed manual we refer to the online manual [2] which is available at

`http://www.dm.unipi.it/~caruso`.

3.1   Loading

The whole library is loaded with `load(gf)`. If Maxima 5.14 is used the additional command `load(gf_root)` is required to load the functions related to modular roots.

3.2   Defining a Finite Field

In order to start up, we must first define our current finite field and its representation as (1). We do this with the `gf_set(p,fx)` command where

- p is the characteristic,
- fx is the monic irreducible polynomial in x generating the field extension.

This command has the following effects that depend on the value of the global variable largefield which is set to true by default:

- the global variables gf_char, gf_exp and gf_irr are set;
- if the global setting largefield is set to true then the smallest primitive element is computed and its value is stored in the global variable pe;
- if the global setting largefield was set to false a complete table of the logarithms and of the powers is computed.

The elements of the field will be represented as polynomials in x.
The command gf_info() prints information about the current field.

```
(%i1) load(gf);
(%o1)                    .../share/contrib/gf/gf.mac
(%i2) gf_set(2,x^4+x+1);
(%o2)                    true
```

3.3   Basic Operations

Addition, subtraction, inversion, multiplication and division are provided by the following commands: gf_add, gf_sub, gf_inv, gf_mul, gf_div.

```
(%i3) gf_mul(x^3+x^2+1,x^3+x+1);
(%o3)                    x² + x
```

$$x^2 + x$$

3.4   Primitive Elements, Powers and Logarithms

The command gf_log(a) computes the discrete logarithm of the a with respect to the primitive element. The command gf_log(f,g) finds the discrete logarithm of f with respect to g by using the Pohlig-Hellman algorithm if largefield is set to true or by using a precomputed table.

```
(%i4) a:x^3+x^2+1;
(%o4)                    x³ + x² + 1
(%i5) gf_log(a);
(%o5)                    13
```

The command gf_findprim() computes the smallest primitive element.
```
(%i6) gf_findprim();
(%o6)                    x
```

The command gf_exp(a,n) computes the n-th power of a:
```
(%i7) ev(a=gf_exp(x,13));
(%o7)                    true
```

3.5   Modular Roots

In order to use the functions related to modular roots in Maxima 5.14, the subpackage gf_roots must be loaded with the load(gf_roots) command. With Maxima 5.15 both parts of the package are loaded automatically.

The functions msqrt and mcbrt, as well as their improved versions, gf_sqrt and gf_cbrt, provide implementations of the modular square and cubic roots:

```
(%i1) load(gf_roots);
(%o1)              .../share/contrib/gf/gf_roots.mac
(%i2) msqrt(100,41);
(%o2)                      [31, 10]
(%i3) mcbrt(343,1789);
(%o3)                    [1064, 718, 7]
(%i4) gf_sqrt(441,11592740641);
(%o4)                  [21, 11592740620]
(%i5) gf_cbrt(1331,11592740641);
(%o5)            [9979650219, 11, 1613090411]
```

4   Applications

The finite fields implementation is robust and efficient enough to deal with the standard applications. For example the Advanced Encryption Standard Rijndael [6] is based on algebra over the finite field

$$\mathbb{F}_2[x]/(x^8 + x^4 + x^3 + x + 1).$$

This field is small enough for the largefield flag to be set to false, so that a tables of powers and logarithms are computed as the field is defined. The "MixColumn" layer of the cryptosystem can be implemented as a matrix product

$$\begin{bmatrix} d_{0,0} & d_{0,1} & d_{0,2} & d_{0,3} \\ d_{1,0} & d_{1,1} & d_{1,2} & d_{1,3} \\ d_{2,0} & d_{2,1} & d_{2,2} & d_{2,3} \\ d_{3,0} & d_{3,1} & d_{3,2} & d_{3,3} \end{bmatrix} = \begin{bmatrix} 2 & 3 & 1 & 1 \\ 1 & 2 & 3 & 1 \\ 1 & 1 & 2 & 3 \\ 3 & 1 & 1 & 2 \end{bmatrix} \begin{bmatrix} c_{0,0} & c_{0,1} & c_{0,2} & c_{0,3} \\ c_{1,0} & c_{1,1} & c_{1,2} & c_{1,3} \\ c_{2,0} & c_{2,1} & c_{2,2} & c_{2,3} \\ c_{3,0} & c_{3,1} & c_{3,2} & c_{3,3} \end{bmatrix}$$

where $c_{i,j}$ are the results of the previous layer. Each of $c_{i,j}$ and $d_{i,j}$ are bytes whose bits may be interpreted as the coefficients of a polynomial element in the field. The constant matrix in the product above may be represented in the field as

$$\begin{bmatrix} x & x+1 & 1 & 1 \\ 1 & x & x+1 & 1 \\ 1 & 1 & x & x+1 \\ x+1 & 1 & 1 & x \end{bmatrix}$$

The inverse can be very quickly determined using the `gf_matinv()` command in the library, and is found to be

$$
\begin{bmatrix}
x^3 + x^2 + x & x^3 + x + 1 & x^3 + x^2 + 1 & x^3 + 1 \\
x^3 + 1 & x^3 + x^2 + x & x^3 + x + 1 & x^3 + x^2 + 1 \\
x^3 + x^2 + 1 & x^3 + 1 & x^3 + x^2 + x & x^3 + x + 1 \\
x^3 + x + 1 & x^3 + x^2 + 1 & x^3 + 1 & x^3 + x^2 + x
\end{bmatrix}
\equiv
\begin{bmatrix}
14 & 11 & 13 & 9 \\
9 & 14 & 11 & 13 \\
13 & 9 & 14 & 11 \\
11 & 13 & 9 & 14
\end{bmatrix}
$$

We can also use the library to investigate the workings of the "ByteSub" layer, which is usually implemented as a lookup table, but which is based on inversion in the field coupled with a linear operation.

For an application of our package to the study of the Chor-Rivest [3] cryptosystem we refer to the on line manual [2].

Since Maxima has a more user-friendly environment and the library uses polynomials in x as input and output, it is very suitable for teaching. Axiom, in comparison, uses elements of the form %A, %B and so on for the output of its field operations. Although this does have the mathematical advantage of allowing different fields to be implemented simultaneously, it is confusing for students. When the third author experimented with teaching a cryptography course using both Axiom and Maxima [12], students found the implementation of Maxima's finite fields easy to use, and intuitive.

5 Performance and Comparisons

When we started working on this library no routines for finite fields were available in Maxima as it still is the case for other free and open source systems, such as GP/Pari [7] for which only recently a patch is being implemented.[4] Axiom provides a "domain", i.e. a type for finite fields with efficient implementations. As well as Axiom, we compare our results with two commercial packages which provide an implementation of finite fields: Maple [8], and MuPAD [5].

The field $\mathbb{F}_2[x]/(x^{20} + x^3 + x^2 + x + 1)$.

Problem	Maxima	Axiom	Maple	MuPAD
Log of $x^{10} + 1$ to the base $x^2 + x$	0.15	≈ 0	≈ 0	0.02
Min. polynomial of $x^{10} + 1$	0.30	≈ 0	—	—

The field $\mathbb{F}_2[x]/(x^{48} + x^{26} + x^{13} + 1)$.

Problem	Maxima	Axiom	Maple	MuPAD
Compute $(x^{10} + 1)^{3^{30}}$	0.15	0.02	0.16	0.04
Random element to the power of 3^{30}	0.35	0.03	≈ 0	0.044

---

[4] Bill Allombert has been working on a patch that adds support for finite fields into GP/Pari. The patch is available at
http://pari.math.u-bordeaux.fr/archives/pari-dev-0703/msg00011.html

The field $\mathbb{F}_7[x]/(x^{10} + 5x^2 + x + 5)$.

Problem	Maxima	Axiom	Maple	MuPAD
Compute $(x^8 + 5x^2 + 3x + 1)^{5^{12}}$	0.05	0.01	≈ 0	0.02
Random element to the power of 5^{12}	0.05	0.01	≈ 0	0.02
Min. polynomial of $x^9 + 3x^6 + x^5 + 2x^2 + 6$	0.16	0.01	–	–
Log of $x^9 + 3x^6 + x^5 + 2x^2 + 6$ in base x	1.15	0.16	–	25.577

Primitive Element

Field	Result	Maxima	Axiom	Ratio	Maple	MuPAD
$\mathbb{F}_{123127}[x]/(x^5 + 2x + 1)$	$x + 4$	0.30	0.18	166.6	0.012	0.112
$\mathbb{F}_{8796519617}[x]/(x^8 + 3x^6 + x + 1)$	$x + 9$	1.39	5.76	24.1	0.344	2.404
$\mathbb{F}_7[x]/(x^{10} + 5x^2 + x + 5)$	x	0.15	0.04	375	≈ 0	0.052
$\mathbb{F}_{32717}[x]/(x^{11} + x^5 + x^2 + x + 1)$	$x + 2$	1.69	5.63	30	0.848	2.376
$\mathbb{F}_{211}[x]/(x^{17} + 2x^2 + 1)$	$x + 6$	0.67	2.12	31.6	0.008	0.908
$\mathbb{F}_2[x]/(x^{20} + x^3 + x^2 + x + 1)$	$x^2 + x$	0.40	0.08	500	≈ 0	0.032
$\mathbb{F}_{197}[x]/(x^{24} - x^8 + 2)$	$x + 19$	1.63	38.73	4.2	0.104	7.897
$\mathbb{F}_5[x]/(x^{61} + x^{15} + 1)$	$x + 4$	1.02	33.98	3	12.448	4.416
$\mathbb{F}_7[x]/(x^{61} + x^4 + 1)$	$x + 3$	0.69	35.53	1.94	1.885	7.976
$\mathbb{F}_5[x]/(x^{84} + x^{41} + x^2 + 1)$	$x^2 + 1$	6.15	323.13	1.9	0.524	38.175
$\mathbb{F}_3[x]/(x^{91} + x^{35} + x + 1)$	x	0.95	39.03	2.43	3.180	5.2
$\mathbb{F}_2[x]/(x^{102} + x^{29} + 1)$	$x + 1$	0.84	34.55	2.43	0.028	4.3

Note: The timings are in seconds. The ratio is the percentage ratio between the Axiom and Maxima timings. The symbol – means that either the system cannot perform the operation or no command for that operation is available in that system.

We focused our efforts on the efficient computation of primitive elements and modular k-th roots for finite fields. The computation of the primitive element in Maxima[5] is generally faster than in Axiom [10] or MuPAD[6] except for some finite fields with very low exponent. We may also note that Maple's timing vary widely, with some of our timings showing a greater speed for Maxima. Considering that Maxima is using interpreted code, while other CAS are generally using compiled code, these figures indicate very good results for our implementation.

6   Future Plans

As already mentioned the upcoming Maxima version 5.15 will include an updated library, providing routines for modular k-th root computation as described in Section 2.6, 2.7 and 2.8.

---

[5] We used Maxima 5.12 with GCL and the latest version of our package, which we tested against Axiom 3.9. The hardware used was an Intel Core 2 E6570 Duo at 2.66Ghz, on x86_64 GNU Linux 2.6.24.

[6] The timings for Maple and MuPAD were done using Maple 10 and MuPAD Pro 3.2 under Linux 2.6.13 on a Pentium IV at 2.66Ghz which is roughly 50% slower than the hardware used for Maxima and Axiom.

As a next step we plan to add new features and further optimizations to our routines. In particular our interest will be focused on

- several optimizations for gf_log and gf_inv (using the Itoh-Tsujii algorithm [9]);
- faster generation and better handling of normal bases (especially for $\mathbb{F}_{2^n}$);
- efficient irreducibility tests over $\mathbb{F}_p$ for polynomials of low degree;
- functions dealing with binary quadratic forms of a given discriminant over $\mathbb{F}_p$;
- cubic and quartic reciprocity symbols (in order to decrease the complexity of tests like $a^{\frac{p-1}{3}} \equiv 1 \pmod{p}$
  from $O(\log(p))$ to $O(\log(a))$ multiplications in $\mathbb{F}_p$);
- implementing optimized routines for $\mathbb{F}_{2^n}$;
- implementing algorithms [20] for the computation of irreducible polynomials over finite fields;
- porting our code to the Axiom system.

Acknowledgments

We would like to thank Carlo Traverso, Dario Bini and Marco Bodrato for their suggestions and encouragement.

References

1. Brickell, E., Gordon, D., McCurley, K., Wilson, D.: Fast exponentiation with precomputation. In: Rueppel, R.A. (ed.) EUROCRYPT 1992. LNCS, vol. 658, pp. 200–207. Springer, Heidelberg (1993)
2. Caruso, F., D'Aurizio, J., Mc Andrew, A.: On line manual on Finite Fields in Maxima (2007), http://www.dm.unipi.it/~caruso
3. Chor, B., Rivest, R.L.: A knapsack-type public key cryptosystem based on arithmetic in finite fields. IEEE Trans. Inform. Theory 34(5, part 1) 901–909 (1988)
4. Cipolla, M.: Sulla risoluzione apiristica delle congruenze binomie secondo un modulo primo. Mathematische Annalen 63, 54–61 (1907)
5. Creutzig, C., Oevel, W.: MuPAD Tutorial, 2nd edn. Springer, Heidelberg (2004)
6. Daemen, J., Rijmen, V.: The design of Rijndael. In: Information Security and Cryptography. AES—the advanced encryption standard. Springer, Heidelberg (2002)
7. Pari Group. GP/Pari on line documentation (2003), http://pari.math.u-bordeaux.fr/
8. Heck, A.: Introduction to Maple, 3rd edn. Springer, Heidelberg (2003)
9. Itoh, T., Tsujii, S.: A fast algorithm for computing multiplicative inverses in $\mathbb{F}_{2^m}$ using normal bases. Inform. and Comput. 78(3), 171–177 (1988)
10. Jenks, R.D., Sutor, R.S.: AXIOM. The scientific computation system, With a foreword by David V. Chudnovsky and Gregory V. Chudnovsky. Numerical Algorithms Group Ltd., Oxford (1992)
11. Lehmer, D.H.: Computer technology applied to the theory of numbers. In: Studies in Number Theory, pp. 117–151; Math. Assoc. Amer. (distributed by Prentice-Hall, Englewood Cliffs, N.J.) (1969)

12. McAndrew, A.: Teaching cryptography with open-source software. In: SIGCSE 2008: Proceedings of the 39th SIGCSE technical symposium on Computer science education, pp. 325–329. ACM, New York (2008)
13. McEliece, R.J.: Finite Fields for Computer Scientists and Engineers. Kluwer Academic Publishers, Boston (1987)
14. Menezes, A.J., van Oorschot, P.C., Vanstone, S.A.: Handbook of applied cryptography. CRC Press Series on Discrete Mathematics and its Applications. CRC Press, Boca Raton (1997) (With a foreword by Ronald L. Rivest)
15. Müller, S.: On the Computation of Square Roots in Finite Fields. Designs, Codes and Cryptography 31(3), 301–312 (2004)
16. Pohlig, S.C., Hellman, M.E.: An improved algorithm for computing logarithms over GF(p) and its cryptographic significance. IEEE Trans. Information Theory IT-24(1), 106–110 (1978)
17. Rowney, K.T., Silverman, R.D.: Finite field manipulations in Macsyma. SIGSAM Bull. 23(1), 39–48 (1989)
18. Schelter, W.F., The Maxima Group: Maxima on line documentation (2001), http://maxima.sourceforge.net
19. Shanks, D.: Five Number-Theoretic Algorithms. In: Proceedings of the Second Manitoba Conference on Numerical Mathematics, pp. 51–70 (1972)
20. Shoup, V.: New Algorithms for Finding Irredicible Polynomials Over Finite Fields. Mathematics of Computation 54(189), 435–447 (1990)
21. Shoup, V.: Searching for Primitive Roots in Finite Fields. Math. Comp. 58(197), 369–380 (1992)
22. Stickelberger, L.: Über eine neue Eigenschaft der Diskriminanten algebraischer Zahlkörper. In: Verhandlungen des ersten Internationalen Mathematiker-Kongresses, pp. 182–193 (1897)
23. Tonelli, A.: Bemerkung über die Auflösung quadratischer Congruenzen. Göttingen Nachrichten, 344–346 (1891)
24. Wang, Y.: On the least primitive root of a prime. Sci. Sinica 10, 1–14 (1961)
25. Yan, S.Y.: Number Theory for Computing, 2nd edn. Springer, New York (2002)

Modular Reduction in $GF(2^n)$ without Pre-computational Phase

M. Knežević[1], K. Sakiyama[1,2], J. Fan[1], and I. Verbauwhede[1]

[1]Katholieke Universiteit Leuven
Department Electrical Engineering - ESAT/SCD-COSIC and IBBT
Kasteelpark Arenberg 10, B-3001 Leuven-Heverlee, Belgium
{mknezevi,ksakiyam,jfan,iverbauw}@esat.kuleuven.be
[2]University of Electro-Communications
Dept. of Information and Communication Eng.
1-5-1 Chofugaoka, Chofu, Tokyo 182-8585, Japan
saki@ice.uec.ac.jp

Abstract. In this study we show how modular multiplication with Barrett and Montgomery reductions over certain finite fields of characteristic 2 can be implemented efficiently without using a pre-computational phase. We extend the set of moduli that is recommended by Standards for Efficient Cryptography (SEC) by defining two distinct sets for which either Barrett or Montgomery reduction is applicable. As the proposed algorithm is very suitable for a fast modular multiplication, we propose an architecture for the fast modular multiplier that can efficiently be used without pre-computing the inverse of the modulus.

Keywords: Modular multiplication, Barrett reduction, Montgomery reduction, elliptic curve cryptography, public-key cryptography.

1 Introduction

Modular multiplication is at the heart of many Public-Key Cryptosystems (PKC), e.g. RSA [10], Diffie-Hellman key agreement [4], ElGamal scheme [5,6] and Elliptic Curve Cryptography (ECC) [7,8]. Due to the very long numbers used in these crypto primitives efficient hardware and software implementation of modular multiplication has always been a challenge. Algorithms that are most commonly used to avoid computationally intensive multi-precision divisions are Barrett reduction [1] and Montgomery reduction [9]. Both algorithms have one common property, namely a pre-computational step, where the inverse of the modulus is calculated and stored together with the value of the modulus. As the modular inverse operation is computationally more expensive than the modular multiplication itself one usually fixes the value of modulus and uses the pre-computed value of the inverse. This reduces flexibility as well as the performance of the implementation increasing the area needed for the storage of the modulus inverse.

In [11] the authors outline a so called unbalanced exponent modular reduction for special type of moduli that can efficiently be used as a replacement for existing

J. von zur Gathen, J.L. Imaña, and Ç.K. Koç (Eds.): WAIFI 2008, LNCS 5130, pp. 77–87, 2008.
© Springer-Verlag Berlin Heidelberg 2008

Barrett and Montgomery algorithms. The original idea comes from Standards for Efficient Cryptography (SEC) [12] and it recommends the use of moduli that are of type $M(x) = x^n + T(x)$ where the degree of $T(x)$ is far smaller than n. This is very suitable for software implementations and makes reduction very efficient due to the special type of modulus. However, hardware implementation still requires more effort, especially when one needs to support more than one modulus.

In this paper we show how Barrett and Montgomery reduction algorithms over binary fields can be performed without using a pre-computational step. We extend the set of moduli that is recommended by SEC by defining two distinct sets for which either Barrett or Montgomery reduction is applicable. Due to the similarity between the Barrett and Montgomery algorithms, we propose a hardware architecture for the fast modular multiplier that can efficiently be used for the two defined sets of moduli. The multiplier performs modular multiplications within a single clock cycle.

The remainder of this paper is structured as follows. Section 2 describes the Barrett and Montgomery modular multiplication algorithms as the two most commonly used reduction methods. In Sect. 3, we show how pre-computation can be omitted in Barrett and Montgomery algorithms over binary fields. Hardware implementation and comparison with related work is given in Sect. 4. Section 5 concludes the paper and gives some guidelines for the future work.

2   Related Work

2.1   Modular Multiplication with Barrett Reduction

Barrett reduction algorithm was introduced by P. D. Barrett in 1986 [1]. This algorithm computes $r \equiv a \bmod m$ for an input a and a modulus m and is given in Alg. 1. The algorithm uses μ, a pre-computed reciprocal of m, to avoid computationally expensive divisions that are necessary to compute the quotient q such that $a = qm + r$.

Modular multiplication of two n-bit inputs using Barrett reduction is done by providing the result of $n \times n$-bit multiplication (input a in Alg. 1).

2.2   Modular Multiplication with Montgomery Reduction

Montgomery algorithm is one of the most commonly used reduction algorithms. In contrast to Barrett reduction it utilizes right to left divisions which make implementation simpler, with no correction steps necessary. The result of reduction has a form $aR^{-1} \bmod m$, where R is a power of the base b. Similar to Barrett reduction this algorithm uses a pre-computed value $\beta \equiv -m^{-1} \bmod R$. Algorithm 2 shows Montgomery reduction in short.

Similar to Barrett reduction, modular multiplication of two n-bit inputs can also be done based on Alg. 2.

Algorithm 1. Barrett reduction for integers

Require: positive integers $a = (a_{2n-1}, ..., a_0)_b$, $m = (m_{n-1}, m_{n-2}..., m_0)_b$ and
$\mu = b^{2n}$ div m, where $m_{n-1} \neq 0$ and $b > 3$.
Ensure: $r \equiv a \bmod m$.
$q_1 \leftarrow a$ div b^{n-1}, $q_2 \leftarrow \mu q_1$, $q_3 \leftarrow q_2$ div b^{n+1}.
$r_1 \leftarrow a \bmod b^{n+1}$, $r_2 \leftarrow mq_3 \bmod b^{n+1}$, $r \leftarrow r_1 - r_2$.
Final reduction and correction step:
if $r \leq 0$ **then**
  $r \leftarrow r + b^{n+1}$.
end if
while $r \geq m$ **do**
  $r \leftarrow r - m$.
end while
return r.

Algorithm 2. Montgomery reduction for integers

Require: positive integers $a = (a_{2n-1}, ..., a_0)_b$, $m = (m_{n-1}..., m_1, m_0)_b$ and
$\beta \equiv -m^{-1} \bmod R$, where $R = b^n$ and $\gcd(b, m) = 1$.
Ensure: $t \equiv aR^{-1} \bmod m$.
$s_1 \leftarrow a \bmod R$, $s_2 \leftarrow \beta s_1 \bmod R$, $s_3 \leftarrow ms_2$.
$t \leftarrow (a + s_3)/R$.
Final reduction:
if $t \geq m$ **then**
  $t \leftarrow t - m$.
end if
return t.

2.3 Shortcomings of the Existing Algorithms

Both described, Barrett and Montgomery algorithms have one property in common. In order to perform modular reduction, they need a pre-computed value of the reciprocal/inverse of modulus. This reduces flexibility of the system forcing us to use fixed modulus and its pre-computed reciprocal/inverse. From the implementation point of view, this requires extra computational time and memory space to store this pre-computed value.

In the next section we show how these shortcomings can be overcome using the special set of moduli over GF(2^n).

3 The Proposed Modular Reduction Method

Here, we first show how the original Barrett reduction can be adapted for modular multiplication over GF(2^n). Second, we provide a special set of moduli for which pre-computational step in Barrett algorithm can be omitted. Finally, we show how Montgomery reduction, using a complementary set of moduli, can also be performed without pre-computing the inverse. Since neither of them requires a pre-computational step these algorithms are specially suitable for both hardware and software implementations.

Before describing the actual algorithms, we need to give some mathematical background of the finite fields arithmetic. Thus, in the next subsection we first give two lemmata and one definition that are necessary for further explanation of the algorithm.

3.1   Mathematical Background

Starting with the basic idea of the proposed Barrett algorithm over $GF(2^n)$ we give Lemma 1 as follows:

Lemma 1. *Let* $M(x) = x^n + \sum_{i=0}^{l} m_i x^i$ *and* $\mu(x) = x^{2n}$ *div* $M(x)$ *be polynomials over* $GF(2)$, *where* $l = \lfloor \frac{n}{2} \rfloor$. *Then it holds:*

$$\mu(x) = M(x) \ . \tag{1}$$

Proof. In order to prove that Eq. (1) holds we need to find polynomial $B(x)$ of degree $n - 1$ or less that satisfies the following equation:

$$x^{2n} = M(x)^2 + B(x) \ .$$

Indeed, if we write x^{2n} as

$$x^{2n} = M(x)^2 + B(x)$$

$$= x^{2n} + \sum_{i=0}^{l} m_i x^{2i} + \sum_{i=0}^{n-1} b_i x^i \ ,$$

we can choose coefficients b_i, $0 \leq i \leq n - 1$, such that $b_{2j} = m_j$ and $b_{2j+1} = 0$, $0 \leq j \leq l$. This concludes the proof.   □

Lemma 2. *Let* $M(x) = \sum_{i=l}^{n} m_i x^i + 1$ *and* $\beta(x) \equiv -M(x)^{-1} \bmod x^n$ *be polynomials over* $GF(2)$, *where* $l = \lceil \frac{n}{2} \rceil$. *Then it holds:*

$$\beta(x) = M(x) \ . \tag{2}$$

Proof. In order to prove Eq. (2) we need to show that

$$M(x)^2 \equiv 1 \bmod x^n \ .$$

Indeed, if we write $M(x)^2$ as

$$M(x)^2 = M(x)M(x)$$

$$= \sum_{i=l}^{n} m_i x^{2i} + 1 \ ,$$

it becomes obvious that $M(x)^2 \equiv 1 \bmod x^n$, since $l = \left\lceil \frac{n}{2} \right\rceil$. This concludes the proof. $\square$

Definition 1. *Let $P(x)$ and $Q(x)$ denote arbitrary polynomials of degree p and q, respectively. We define $\Delta(n) = \frac{P(x)}{Q(x)}$ such that $n = p - q$ and $n \in \mathbb{Z}$. In other words, with $\Delta(n)$ we denote an arbitrary element from the set of all rational functions of degree n.*

3.2   Barrett Reduction without Pre-computation

The Barrett modular reduction algorithm for integers is given in Sect. 2. In [3] the author shows how the original Barrett reduction can be adapted for the finite fields of characteristic q. Here, we outline the Barrett reduction over a binary field and additionally, we propose a special set of moduli for which the pre-computational step is not needed. First, we provide Alg. 3 and then we give a proof.

Algorithm 3. Barrett reduction over GF(2^n)

Require: polynomial-basis inputs $A(x) = \sum_{i=0}^{2n} a_i x^i$, $M(x) = x^n + \sum_{i=0}^{n-1} m_i x^i$ and $\mu(x) = x^{2n} \operatorname{div} M(x)$, where $a_i, m_i \in \{0, 1\}$.
Ensure: $R(x) \equiv A(x) \bmod M(x)$.
  $Q_1(x) \leftarrow A(x) \operatorname{div} x^n$, $Q_2(x) \leftarrow \mu(x)Q_1(x)$, $Q_3(x) \leftarrow Q_2(x) \operatorname{div} x^n$.
  $R_1(x) \leftarrow A(x) \bmod x^n$, $R_2(x) \leftarrow M(x)Q_3(x) \bmod x^n$, $R(x) \leftarrow R_1(x) + R_2(x)$.

Proof. Using notation from Def. 1 and starting from the original Barrett reduction algorithm we can write

$$\mu(x) = x^{2n} \operatorname{div} M(x)$$
$$= \frac{x^{2n}}{M(x)} + \Delta(-1) \ .$$

Similarly, we can express $Q_1(x)$ as

$$Q_1(x) = A(x) \operatorname{div} x^n$$
$$= \frac{A(x)}{x^n} + \Delta(-1)$$
$$= \left(\frac{Q(x)M(x)}{x^n} + \Delta(-1) \right) + \Delta(-1)$$
$$= \frac{Q(x)M(x)}{x^n} + \Delta(-1) \ ,$$

where $Q(x) = A(x) \operatorname{div} M(x)$. Using the previous equations, $Q_2(x)$ and $Q_3(x)$ can be written as

$$
\begin{aligned}
Q_2(x) &= \mu(x)Q_1(x) \\
&= \left(\frac{x^{2n}}{M(x)} + \Delta(-1)\right)\left(\frac{Q(x)M(x)}{x^n} + \Delta(-1)\right) \\
&= Q(x)x^n + \frac{x^{2n}}{M(x)}\Delta(-1) + \frac{Q(x)M(x)}{x^n}\Delta(-1) + \Delta(-1)\Delta(-1) \\
&= Q(x)x^n + \Delta(n-1) + \Delta(n-1) + \Delta(-2) \\
&= Q(x)x^n + \Delta(n-1) \; , \\
Q_3(x) &= Q_2(x) \operatorname{div} x^n \\
&= \left(Q(x)x^n + \Delta(n-1)\right) \operatorname{div} x^n \\
&= Q(x) \; .
\end{aligned}
$$

Finally, we can evaluate $R(x) = A(x) \bmod M(x)$ as

$$
\begin{aligned}
R(x) &\equiv A(x) \bmod x^n + M(x)\big(A(x) \operatorname{div} M(x)\big) \bmod x^n \\
&\equiv A(x) \bmod x^n + M(x)Q(x) \bmod x^n \\
&\equiv A(x) \bmod x^n + M(x)Q_3(x) \bmod x^n \; .
\end{aligned}
$$

This concludes the proof. $\qquad\square$

Now, according to Lemma 1, we can define a set of moduli for which the Barrett reduction described in Alg. 3 does not require a pre-computational step. This set is of type $M(x) = x^n + \sum_{i=0}^{l} m_i x^i$, where $l = \lfloor \frac{n}{2} \rfloor$ and the algorithm is shown in Alg. 4. It is interesting to note here that, for this special case, the irreducible polynomial can be chosen from the set that contains $2^{\lfloor n/2 \rfloor}$ different polynomials. As we already know, only irreducible polynomials can be used to construct the field.

Algorithm 4. Barrett reduction over $GF(2^n)$ without pre-computation

---

Require: polynomial-basis inputs $A(x) = \sum_{i=0}^{2n} a_i x^i$, $M(x) = x^n + \sum_{i=0}^{l} m_i x^i$, where $l = \lfloor \frac{n}{2} \rfloor$ and $a_i, m_i \in \{0, 1\}$.

Ensure: $R(x) \equiv A(x) \bmod M(x)$.

$\quad Q_1(x) \leftarrow A(x) \operatorname{div} x^n$, $\quad Q_2(x) \leftarrow M(x)Q_1(x)$, $\quad Q_3(x) \leftarrow Q_2(x) \operatorname{div} x^n$.

$\quad R_1(x) \leftarrow A(x) \bmod x^n$, $\quad R_2(x) \leftarrow M(x)Q_3(x) \bmod x^n$, $\quad R(x) \leftarrow R_1(x) + R_2(x)$.

3.3 Montgomery Reduction without Pre-computation

Since there is no correction step in the original Montgomery algorithm (see Alg. 2), this method can be easily applied for the modular multiplication over GF(2^n) (see Alg. 5). This algorithm was proposed in [2]. Here we outline the algorithm and then, to make the paper more consistent, we also give a proof.

Algorithm 5. Montgomery reduction over GF(2^n)

Require: polynomial-basis inputs $A(x) = \sum_{i=0}^{2n} a_i x^i$, $M(x) = x^n + \sum_{i=1}^{n-1} m_i x^i + 1$ and
$\beta(x) \equiv -M^{-1}(x) \bmod R(x)$, where $R(x) = x^n$ and $a_i, m_i \in \{0, 1\}$.
Ensure: $T(x) \equiv A(x)R(x)^{-1} \bmod M(x)$.
  $S_1(x) \leftarrow A(x) \bmod R(x)$, $S_2(x) \leftarrow \beta(x)S_1(x) \bmod R(x)$, $S_3(x) \leftarrow M(x)S_2(x)$.
  $T(x) \leftarrow \big(A(x) + S_3(x)\big)/R(x)$.
  return $T(x)$.

Proof. Polynomial $A(x)$ can be written as $A(x) = A_1(x)R(x) + A_0(x)$, where $R(x) = x^n$. There exists polynomial $S(x)$ of degree $n-1$ such that

$$A_0(x) + M(x)S(x) \equiv 0 \bmod R(x) \ ,$$

In other words, $S(x)$ can be expressed as

$$S(x) \equiv -A_0(x)M(x)^{-1} \bmod R(x) \ .$$

At the same time it holds

$$A(x) + M(x)S(x) \equiv A(x) \bmod M(x) \ ,$$
$$A(x) + M(x)S(x) \equiv 0 \bmod R(x) \ .$$

Finally, we have

$$T(x) \equiv \big(A(x) + M(x)S(x)\big)/R(x)$$
$$\equiv A(x)R(x)^{-1} \bmod M(x) \ .$$

Using notations from Alg. 5 it is obvious that:

$$\beta(x) = -M^{-1}(x) \bmod R(x)$$
$$S_1(x) = A_0(x)$$
$$S_2(x) = S(x)$$
$$S_3(x) = M(x)S(x) \ .$$

This concludes the proof. □

According to Lemma 2, we can easily find a set of moduli for which the pre-computational step in Montgomery reduction can be omitted. Instead of pre-computing and using $\beta(x)$ we use the modulus itself. This algorithm is shown in Alg.6. The proposed set is of type $M(x) = x^n + \sum_{i=l}^{n-1} m_i x^i + 1$, where $l = \lceil \frac{n}{2} \rceil$. Similar to the set defined for Barrett reduction, this set also contains $2^{\lfloor n/2 \rfloor}$ different polynomials.

Algorithm 6. Montgomery reduction over $\mathrm{GF}(2^n)$ without pre-computation

---

Require: polynomial-basis inputs $A(x) = \sum_{i=0}^{2n} a_i x^i$ and $M(x) = x^n + \sum_{i=l}^{n-1} m_i x^i + 1$

where $l = \lceil \frac{n}{2} \rceil$, $R(x) = x^n$ and $a_i, m_i \in \{0, 1\}$.

Ensure: $T(x) \equiv A(x) R^{-1}(x) \bmod M(x)$.

$S_1(x) \leftarrow A(x) \bmod R(x)$, $S_2(x) \leftarrow M(x) R_1(x) \bmod R(x)$,

$S_3(x) \leftarrow M(x) S_2(x)$.

$T(x) \leftarrow \big(A(x) + S_3(x)\big) \operatorname{div} R(x)$.

return $T(x)$.

---

4   Hardware Implementation of the Proposed Algorithm

To verify our algorithm in practice we synthesize the proposed solution for $\mathrm{GF}(2^{192})$ using a $0.13\mu m$ CMOS standard cell library. Here, we aim only for the fast version of the multiplier. Additionally, we synthesize standard Barrett and Montgomery reduction algorithms and compare them with our results. To make a fair comparison we use the same 192×192-bit multipliers in every implementation.

An architecture for the straightforward implementation of the Barrett or Montgomery modular multiplication algorithm over $\mathrm{GF}(2^{192})$ is shown in Fig. 1 (a). For both algorithms we need five 192-bit registers, three 192×192-bit multipliers and one 192-bit adder. Since in binary fields there is no carry propagation, addition is equivalent to XOR operation.

A block diagram of the proposed solution is shown in Fig. 1 (b). Here we can see that instead of using five we use only four 192-bit registers. Similarly to Barrett and Montgomery architecture shown in Fig. 1 (a) we use three 192×192-bit multipliers and one 192-bit adder. Additionally, we use four multiplexers that are driven by the register *ind*. This 1-bit register indicates which of the two proposed complementary set of moduli is used. Architecture of the selectors 1, 2 and 3 is given in Fig. 2. For *ind* $= 0$ our architecture executes Barrett modular multiplication while for *ind* $= 1$ it performs Montgomery modular multiplication.

Synthesis results are given in Table 1. They include registers and combinational logic. For both Barrett and Montgomery algorithms, we assume that

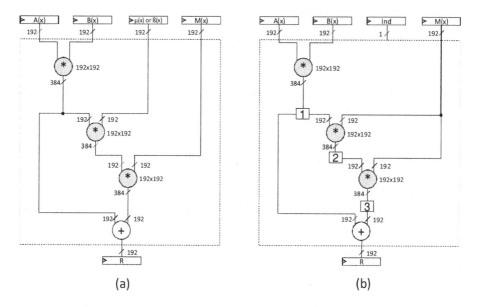

Fig. 1. Architecture of the modular multiplier over GF(2^{192}). (a) using standard Barrett or Montgomery reduction. (b) using proposed method.

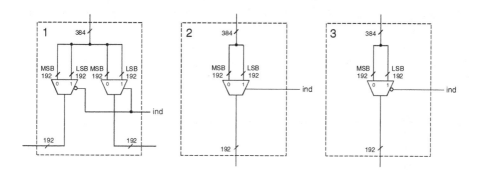

Fig. 2. Architecture of the three selectors in proposed modular multiplier

Table 1. Comparison of the hardware implementations for modular multiplication over GF(2^{192})

Architecture	Pre-computation needed	Size [kgate]	Latency [ns]	Number of cycles
Barrett*	Yes	405.61	6.12	1
Montgomery*	Yes	405.61	6.12	1
Proposed	No	406.32	6.69	1

*Area and latency for performing the pre-computational step is not included.

pre-computation is already performed and both values for $\mu(x)$ and $\beta(x)$ (see Fig. 1 (a)) are known. Skipping the pre-computation step is highly beneficial for both area and computational cost and is the main advantage of our algorithm.

Observing the results from Table 1 we can conclude that our architecture has almost identical size as the separate Barrett and Montgomery multipliers. This gives a practical value to our theoretical work described above. One can always use the proposed implementation with much more flexibility and choose different moduli without pre-computational phase.

5 Conclusions and Future Work

In this paper we have defined two distinct sets of moduli for which the pre-computational step in the modular multiplication algorithm can be excluded. Sets are of type $M(x) = x^n + \sum_{i=0}^{l} m_i x^i$, where $l = \lfloor \frac{n}{2} \rfloor$ and $M(x) = x^n + \sum_{i=l}^{n-1} m_i x^i + 1$, where $l = \lceil \frac{n}{2} \rceil$. Additionally, we have introduced a hardware architecture for the fast modular multiplier over $GF(2^n)$ that uses proposed sets of moduli without pre-computation. Architecture supports both, Barrett and Montgomery modular reduction.

At the cost of more control logic, a similar multiplier that supports different degrees of moduli can be introduced. This would further increase the flexibility of the proposed fast modular multiplier.

As a part of our future research we would like to explore the impact of the similar special sets of moduli on building a compact version of the modular multiplier.

Acknowledgment

We would like to thank Dr. Frederik Vercauteren for answering patiently all our questions and providing useful feedback and comments.

This work is funded partially by IBBT, Katholieke Universiteit Leuven (OT/06/40) and FWO projects (G.0300.07 and G.0450.04). This work was supported in part by the IAP Programme P6/26 BCRYPT of the Belgian State (Belgian Science Policy), by the EU IST FP6 projects (ECRYPT) and by the IBBT-QoE project of the IBBT.

References

1. Barrett, P.: Implementing the Rivest Shamir and Adleman Public Key Encryption Algorithm on a Standard Digital Signal Processor. In: Odlyzko, A.M. (ed.) CRYPTO 1986. LNCS, vol. 263, pp. 311–323. Springer, Heidelberg (1987)
2. Koç, Ç.K., Acar, T.: Montgomery multiplication in $GF(2^k)$. Designs, Codes and Cryptography 14, 57–69 (1998)

3. Dhem, J.-F.: Efficient modular reduction algorithm in $\mathbb{F}_q[x]$ and its application to left to right modular multiplication in $\mathbb{F}_2[x]$. In: Proceedings of 5th International Workshop on Cryptographic Hardware and Embedded Systems (CHES). LNCS, pp. 203–213. Springer, Heidelberg (2003)

4. Diffie, W., Hellman, M.E.: New directions in cryptography. IEEE Transactions on Information Theory 22, 644–654 (1976)

5. ElGamal, T.: A public key cryptosystem and a signature scheme based on discrete logarithms. In: Blakely, G.R., Chaum, D. (eds.) CRYPTO 1984. LNCS, vol. 196, pp. 10–18. Springer, Heidelberg (1985)

6. ElGamal, T.: A public key cryptosystem and a signature scheme based on discrete logarithms. IEEE Transactions on Information Theory 31, 469–472 (1985)

7. Koblitz, N.: Elliptic curve cryptosystem. Math. Comp. 48, 203–209 (1987)

8. Miller, V.: Uses of elliptic curves in cryptography. In: Williams, H.C. (ed.) Advances in Cryptology: Proceedings of CRYPTO 1985. LNCS, vol. 218, pp. 417–426. Springer, Heidelberg (1986)

9. Montgomery, P.: Modular multiplication without trial division. Mathematics of Computation 44(170), 519–521 (1985)

10. Rivest, R.L., Shamir, A., Adleman, L.: A method for obtaining digital signatures and public-key cryptosystems. Communications of the ACM 21(2), 120–126 (1978)

11. Shen, H., Jin, Y., You, R.: Unbalanced Exponent Modular Reduction over Binary Field and Its Implementation. In: Proceedings of the First International Conference on Innovative Computing, Information and Control, pp. 190–193 (2006)

12. Standards for Efficient Cryptography. Elliptic Curve Cryptography, Version 1.5, draft (2005), http://www.secg.org

Subquadratic Space Complexity Multiplication over Binary Fields with Dickson Polynomial Representation

M. Anwar Hasan[1] and Christophe Negre[2]

[1] Department of Electrical and Computer Engineering, University of Waterloo, Canada
[2] Team DALI/ELIAUS, University of Perpignan, France

Abstract. We study Dickson bases for binary field representation. Such a representation seems interesting when no optimal normal basis exists for the field. We express the product of two elements as Toeplitz or Hankel matrix vector product. This provides a parallel multiplier which is subquadratic in space and logarithmic in time.

1 Introduction

Finite field arithmetic is extensively used in cryptography. For public key cryptosystems, the size (i.e. the number of element) of the field may be quite large, say 2^{2048}. Finite field multiplication over such a large field requires a considerable amount of resources (time or space). For binary extension fields, used in many practical public key cryptosystems, field elements can be represented with respect to a normal basis, where squaring operations are almost free of cost. In order to reduce the cost of multiplication over the extension field, instead of using an arbitrary normal basis, it is desirable to use an optimal normal basis. The latter however does not exist for all extension fields, in which case one may use Dickson bases [2,7] and develop an efficient field multiplier.

In this paper we consider subquadratic space complexity multipliers using the Dickson basis. To this end, using low weight Dickson polynomials, we formulate the problem of field multiplication as a product of a Toeplitz or Hankel matrix and a vector, and apply subquadratic space complexity algorithm for the product [4], which gives us a subquadratic space complexity field multiplier.

The article is organized as follows. In Section 2 we present some general results on Dickson polynomials. In Section 3 we give the outline of the subquadratic multiplier of matrix vector product of [4]. Then in Section 4 we give a matrix vector product approach in Dickson basis representation. We wind up with complexity comparison and a brief conclusion.

2 Dickson Polynomials

Dickson polynomials over finite fields were introduced by L.E. Dickson in [2]. These polynomials have several applications and interesting properties, the main

J. von zur Gathen, J.L. Imaña, and Ç.K. Koç (Eds.): WAIFI 2008, LNCS 5130, pp. 88–102, 2008.

one being a permutation property over finite fields. For a complete explanation on this the reader may refer to [6]. Our interest here concerns the use of Dickson polynomial for finite field representation for efficient binary field multiplication. There are two kinds of Dickson polynomials, and there are several ways to define and construct both of them. We give here the definition of [6] of the first kind Dickson polynomials.

Definition 1 (Dickson Polynomial[6] page 9). *Let R be a ring and $a \in R$. The Dickson polynomial of the first kind $D_n(X, a)$ is defined by*

$$D_n(X, a) = \sum_{i=0}^{\lfloor n/2 \rfloor} \frac{n}{n-i} \binom{n-i}{n} (-a)^i X^{n-2i}. \tag{1}$$

For $n = 0$, we set $D_0(X, a) = 2$ and for $n = 1$ we have $D_1(X, a) = X$.

In [6] it has been shown that Dickson polynomials can be computed using the following recursive relation

$$\begin{cases} D_0(X, a) = 2, \\ D_1(X, a) = X, \\ D_n(X, a) = X D_{n-1}(X, a) - a D_{n-2}. \end{cases} \tag{2}$$

Using these relations we obtain the Dickson polynomials $D_n(X, 1)$ in $\mathbb{F}_2[X]$ for $n \leq 20$ given in Table 1.

The following theorem will be extensively used for the construction of subquadratic multipliers in the Dickson basis.

Table 1. Dickson polynomials

β_1	X
β_2	X^2
β_3	$X^3 + X$
β_4	X^4
β_5	$X^5 + X^3 + X$
β_6	$X^6 + X^2$
β_7	$X^7 + X^5 + X$
β_8	X^8
β_9	$X^9 + X^7 + X^5 + X$
β_{10}	$X^{10} + X^6 + X^2$
β_{11}	$X^{11} + X^9 + X^5 + X^3 + X$
β_{12}	$X^{12} + X^4$
β_{13}	$X^{13} + X^{11} + X^9 + X^3 + X$
β_{14}	$X^{14} + X^{10} + X^2$
β_{15}	$X^{15} + X^{13} + X^9 + X$
β_{16}	X^{16}
β_{17}	$X^{17} + X^{15} + X^{13} + X^9 + X$
β_{18}	$X^{18} + X^{14} + X^{10} + X^2$

Theorem 1. *We denote $\beta_i = D_i(X, 1)$ the n-th Dickson polynomial in $\mathbb{F}_2[X]$. Then for all $i, j \geq 0$ the following equation holds*

$$\beta_i \beta_j = \beta_{i+j} + \beta_{|i-j|}. \tag{3}$$

Proof (Proof). This theorem is a consequence of equation (2).

We will show it by induction on i and j. Using Table 1 We can easily check that equation (3) holds for $i, j \leq 1$. We suppose that the equation is true for all $i, j \leq n$ and we prove that the equation is true for $i, j \leq n + 1$. We first prove it for $i = n + 1$ and $j \leq n$. We have

$$\beta_{n+1}\beta_j = (X\beta_n + \beta_{n-1})\beta_j$$
$$= X\beta_n\beta_j + \beta_{n-1}\beta_j = X(\beta_{n+j} + \beta_{|n-j|}) + (\beta_{n-1+j} + \beta_{|n-1-j|}),$$

by induction hypothesis. Now we have

$$\beta_{n+1}\beta_j = (X\beta_{n+j} + \beta_{n+j-1}) + (X\beta_{|n-j|} + \beta_{|n-1-j|})$$
$$= \beta_{n+1+j} + \beta_{|n+1-j|}.$$

For the other case $i = n + 1$ and $j = n + 1$, the product $\beta_{n+1}\beta_{n+1}$ is obtained using similar tricks.

Polynomial and finite field representation using Dickson polynomials. A direct consequence of Definition 1 is that each β_i for $i \geq 1$ has degree i (in equation 1 look at the term corresponding to $i = 0$). As a result each polynomial $A = \sum_{i=0}^{n} A_i X^i \in \mathbb{F}_2[X]$ can be expressed as

$$A = a_0 + \sum_{i=1}^{n} a_i \beta_i.$$

Such expression can be obtained using Algorithm 1.

Algorithm 1. Conversion of field element representation

Require: A polynomial $A(X) \in \mathbb{F}_2[X]$ of degree n.
  $R \leftarrow A$
  for $i = n$ **to** 1 **do**
    if $\deg R = i$ **then**
      $a_i \leftarrow 1$
      $R \leftarrow R + \beta_i$
    else
      $a_i \leftarrow 0$
    end if
  end for
  $a_0 \leftarrow R$
Ensure: Return $(a_0, \ldots, a_n)$

For example for $A = 1 + X^2 + X^5$ the execution of the previous algorithm gives

$R \leftarrow 1 + X^2 + X^5$	
begin for	
$i = 5$	$R \leftarrow R + \beta_5 = 1 + X + X^2 + X^3, a_5 \leftarrow 1$
$i = 4$	$a_4 \leftarrow 0$
$i = 3$	$R \leftarrow R + \beta_3 = 1 + X^2, a_3 \leftarrow 1$
$i = 2$	$R \leftarrow R + \beta_2 = 1, a_2 \leftarrow 1$
$i = 1$	$a_1 \leftarrow 0$
end for	
	$a_0 \leftarrow 1$
$A = 1 + \beta_2 + \beta_3 + \beta_5$	

Since each polynomial can be written in term of Dickson polynomials, we can use Dickson polynomials for basis representation of binary fields.

Remark 1. Algorithm 1 is only a simple version of the conversion process from standard polynomial to Dickson representation. This conversion can be expressed through an $n \times n$ matrix vector product, and can be implemented with a parallel circuit with at most n^2 AND gates and $n(n-1)$ XOR gates with logarithmic time delay. We expect that some Toeplitz structure could appear in the conversion matrix, and the corresponding complexity would be subquadratic.

Theorem 2. *Let P be an irreducible polynomial of degree n in $\mathbb{F}_2[X]$. The system $\mathcal{B} = \{\beta_1, \ldots, \beta_n\}$ forms a basis of $\mathbb{F}_{2^n} = \mathbb{F}_2[X]/(P)$ over $\mathbb{F}_2$.*

Proof (Proof). To show that $\mathcal{B}$ is a basis we have to show that each element $A \in \mathbb{F}_{2^n}$ can be expressed as

$$A = \sum_{i=1}^{n} a_i \beta_i \text{ with } a_i \in \{0,1\},$$

and this expression is unique.

Let us first show that for each $A \in \mathbb{F}_{2^n}$ an expression in $\mathcal{B}$ exists. The polynomial P is an irreducible polynomial in $\mathbb{F}_2[X]$ and using Algorithm 1 it can be expressed as

$$P = 1 + \sum_{i=1}^{n-1} p_i \beta_i + \beta_n.$$

Let $A \in \mathbb{F}[X]/(P)$ which is a polynomial of degree less than n and can also be written as $A = a_0 + \sum_{i=1}^{n-1} a_i \beta_i$ with Algorithm 1. To get required expression of A in $\mathcal{B}$ we need to express the coefficient a_0 in $\mathcal{B}$. To do this, we use the expression of P in $\mathcal{B}$. Since $1 = \sum_{i=1}^{n-1} p_i \beta_i + \beta_n \mod P$ we can replace a_0 by $\sum_{i=1}^{n-1} a_0 p_i \beta_i + a_0 \beta_n$. We finally obtain

$$A = \sum_{i=1}^{n-1} (a_i + a_0 p_i) \beta_i + a_0 \beta_n \mod P.$$

Now we show that such expression is unique. If we have a second different expression $A = \sum_{i=1}^{n} a_i'\beta_i$, then by adding the two we get

$$\sum_{i=1}^{n}(a_i + a_i')\beta_i = 0. \tag{4}$$

Let d be the maximal subscript such that $a_d \neq a_d'$. We rewrite $\beta_d = X^d + \beta_d'$ where $\deg \beta_d' < d$ and then using (4) we obtain

$$\sum_{i=1}^{d-1}(a_i + a_i')\beta_i + (a_d + a_d')\beta_d' + (a_d + a_d')X^d = 0.$$

Now $\deg(\sum_{i=1}^{d-1}(a_i + a_i')\beta_i + (a_d + a_d')\beta_d') \leq d - 1$, and thus we must have $(a_d + a_d')X^d = 0$, this contradicts the fact that $a_d \neq a_d'$.

3 Asymptotic Complexities of Toeplitz Matrix Vector Product

In this section we recall some basics matrix-vector multiplication and their corresponding space and time complexities [4]. A Toeplitz matrix is defined as

Definition 2. *An $n \times n$ Toeplitz matrix is a matrix $[t_{i,j}]_{0 \leq i,j \leq n-1}$ such that $t_{i,j} = t_{i-1,j-1}$ for $1 \leq i, j$.*

If $2|n$ we can use a *two way approach* presented in Table 2, to compute a matrix vector product $T \cdot V$ where T is an $n \times n$ Toeplitz matrix. If $3|n$ we can use the *three way* approach which is also presented in Table 2.

If n is a power of 2 or a power of 3 the formulas of Table 2 can be used recursively to perform $T \cdot V$. Using these recursive processes through parallel computation, the resulting multipliers [4] have the complexity given in Table 3.

The above subquadratic approach can also be used when H is an Hankel matrix. We recall the definition of an Hankel matrix.

Definition 3 (Hankel matrix). *Let $H = [h_{i,j}]_{0 \leq i,j \leq n-1}$ be an $n \times n$ matrix. We say that H is Hankel if*

$$h_{i,j} = h_{i-1,j+1} \text{ for } 1 \leq i \text{ and } j < n - 1 \tag{5}$$

Moreover we say that H is an essentially Hankel matrix, if H satisfies (5) unless for $i = n - 1$ and for $0 \leq j \leq n - 1$ where $h_{n-1,j} = 0$.

Let H be an Hankel matrix. The matrix H' with the same rows as H in the reverse order

$$H' = [h_{n-1-i,j}]_{0 \leq i,j \leq n-1}$$

is a Toeplitz matrix. Consequently to perform $W = H \cdot V$, we compute $W' = H' \cdot V$ using the subquadratic of Table 3 method and then reverse the order of the coefficients of W' to get W.

Table 2. Subquadratic Toeplitz matrix vector product

Matrix decomposition	
Two way	Three way
$T = \begin{bmatrix} T_1 & T_0 \\ T_2 & T_1 \end{bmatrix} \begin{bmatrix} V_0 \\ V_1 \end{bmatrix}$	$T = \begin{bmatrix} T_2 & T_1 & T_0 \\ T_3 & T_2 & T_1 \\ T_4 & T_3 & T_2 \end{bmatrix} \begin{bmatrix} V_0 \\ V_1 \\ V_2 \end{bmatrix}$
Recursive formulas	
$T \cdot V = \begin{bmatrix} P_0 + P_2 \\ P_1 + P_2 \end{bmatrix}$	$T \cdot V = \begin{bmatrix} P_0 + P_3 + P_4 \\ P_1 + P_3 + P_5 \\ P_2 + P_4 + P_5 \end{bmatrix}$
where	where
$P_0 = (T_0 + T_1)V_1,$ $P_1 = (T_1 + T_2)V_0,$ $P_2 = T_1(V_0 + V_1),$	$P_0 = (T_0 + T_1 + T_2)V_2,$ $P_1 = (T_0 + T_1 + T_3)V_1,$ $P_2 = (T_2 + T_3 + T_4)V_0,$ $P_3 = T_1(V_1 + V_2,)$ $P_4 = T_2(V_0 + V_2),$ $P_5 = T_3(V_0 + V_1),$

Table 3. Asymptotic complexity

	Two-way split method	Three-way split method
# AND	$n^{\log_2(3)}$	$n^{\log_3(6)}$
# XOR	$5.5n^{\log_2(3)} - 6n + 0.5$	$\frac{24}{5}n^{\log_3(6)} - 5n + \frac{1}{5}$
Delay	$T_A + 2\log_2(n)T_X$	$T_A + 3\log_3(n)T_X$

4 Field Multiplication Using Low Weight Dickson Polynomials

In this section we consider multiplication of two elements of the binary field $\mathbb{F}_{2^n} = \mathbb{F}_2[X]/(P)$ where the polynomial P is a low weight Dickson polynomial. In particular we consider two and three-term Dickson polynomials P, i.e., Dickson binomials and trinomials. Like low weight conventional polynomials the use of low weight Dickson polynomials is expected to yield lower space complexity multipliers.

4.1 Irreducible Dickson Binomials

In this subsection we will focus on finite fields $\mathbb{F}_{2^n} = \mathbb{F}_2[X]/(P)$ where P is a two terms irreducible polynomial of the form $P = \beta_n + 1$ where β_n is the n-th Dickson polynomial. As shown in Appendix A, for cryptographic size there is no $\beta_n + 1$ irreducible. But some of them admit a *big* irreducible factor P which can be use to define the field $\mathbb{F}_{2^n}$.

The elements of $\mathbb{F}_{2^n}$ are expressed in the Dickson basis $\mathcal{B} = \{\beta_1, \ldots, \beta_n\}$. The following theorem shows that the product of two elements A and B in $\mathbb{F}_{2^n}$ can be computed as a matrix-vector product $M_A \cdot B$ where M_A is a sum of a Toeplitz matrix and an essentially Hankel matrix.

Theorem 3. *Let n be an integer such that $\beta_n + 1$ is irreducible and let $\mathbb{F}_{2^n} = \mathbb{F}_2[X]/(\beta_n + 1)$. Let $A = \sum_{i=1}^{n} a_i\beta_i$ and $B = \sum_{i=1}^{n} b_i\beta_i$ be two elements of $\mathbb{F}_{2^n}$ expressed in $\mathcal{B}$. The coefficients in $\mathcal{B}$ of the product $A \times B$ can be computed as*

$$
\begin{bmatrix}
a_n & a_{n-1}+a_1 & \cdots & a_2+a_{n-2} & a_1+a_{n-1} \\
a_1 & a_n & \cdots & a_3+a_{n-3} & a_2+a_{n-2} \\
\vdots & & & & \vdots \\
a_{n-2} & \cdots & \cdots & a_n & a_{n-1}+a_1 \\
a_{n-1} & \cdots & \cdots & a_1 & a_n
\end{bmatrix}
\cdot
\begin{bmatrix} b_1 \\ \vdots \\ b_n \end{bmatrix}
$$

$$
+
\begin{bmatrix}
a_2 & a_3 & \cdots & a_{n-1} & 0 & a_{n-1} \\
a_3 & a_4 & \cdots & 0 & a_{n-1} & a_{n-2} \\
\vdots & & & & \vdots & \\
0 & a_{n-1} & & & a_2 & a_1 \\
0 & & & & 0 & 0
\end{bmatrix}
\cdot
\begin{bmatrix} b_1 \\ \vdots \\ b_n \end{bmatrix}.
$$

Proof (Proof). If we multiply the two elements A and B we get the following:

$$
AB = \left(\sum_{i=1}^{n} a_i\beta_i \right) \times \left(\sum_{i=1}^{n} b_i\beta_i \right) = \sum_{i,j=1}^{n} a_ib_i\beta_i\beta_j. \tag{6}
$$

Then from Theorem 1 we have $\beta_i\beta_j = \beta_{i+j} + \beta_{|i-j|}$, we can rewrite (6) as

$$
AB = \underbrace{\left(\sum_{i,j=1}^{n} a_ib_j\beta_{i+j} \right)}_{S_1} + \underbrace{\left(\sum_{i,j=1}^{n} a_ib_j\beta_{|i-j|} \right)}_{S_2}
$$

Now we express this former expression of AB as a sum of Toeplitz or Hankel matrix vector product.

Let us begin with S_1. We remark that S_1 has a similar expression as product of two polynomials of the same degree. In other words, S_1 can be computed as $Z_A \cdot B$ where

$$
Z_A =
\begin{bmatrix}
0 & 0 & \cdots & 0 & 0 \\
a_1 & 0 & \cdots & 0 & 0 \\
\vdots & & & & \vdots \\
a_{n-1} & \cdots & \cdots & a_1 & 0 \\
a_n & \cdots & \cdots & a_2 & a_1 \\
0 & a_n & \cdots & a_3 & a_1 \\
\vdots & & & & \vdots \\
0 & 0 & \cdots & 0 & a_n
\end{bmatrix}
\begin{matrix}
\leftarrow \beta_1 \\
\leftarrow \beta_2 \\
\vdots \\
\leftarrow \beta_n \\
\leftarrow \beta_{n+1} \\
\leftarrow \beta_{n+2} \\
\vdots \\
\leftarrow \beta_{2n}
\end{matrix}
$$

We reduce the matrix Z_A modulo $P = \beta_n + 1$ to get non-zero coefficients only on rows corresponding to $\beta_1, \ldots, \beta_n$. We use the fact that β_{n+i} for $i \geq 0$ satisfies

$$\beta_{n+i} = \beta_i \beta_n + \beta_{n-i} = \beta_i + \beta_{n-i}.$$

This equation is a simple consequence of equation (3) and that $\beta_n = 1 \mod P$. This implies that the rows corresponding to β_{n+i} are reduced into two rows one corresponding to β_i and the other to β_{n-i}. After performing this reduction and removing zero rows we get

$$S_1 = Z_A \cdot B = \underbrace{\begin{bmatrix} a_n & a_{n-1} & \cdots & a_2 & a_1 \\ a_1 & a_n & \cdots & a_3 & a_2 \\ & \vdots & & & \vdots \\ a_{n-1} & \cdots & & \cdots & a_1 & a_n \end{bmatrix} \begin{bmatrix} b_1 \\ \vdots \\ b_n \end{bmatrix}}_{S_{1,1}}$$

$$+ \underbrace{\begin{bmatrix} 0 & 0 & \cdots & a_n & a_{n-1} \\ \vdots & & & & \vdots \\ 0 & a_n & \cdots & a_3 & a_1 \\ a_n & a_{n-1} & \cdots & a_2 & a_1 \\ 0 & \cdots & & \cdots & 0 & 0 \end{bmatrix} \begin{bmatrix} b_1 \\ \vdots \\ b_n \end{bmatrix}}_{S_{1,2}}$$

Finally, we get an expression of S_1 as matrix vector product where the matrix is a sum of a Toeplitz and an essentially Hankel matrix.

Now we do the same for S_2. We split S_2 into two sums

$$S_2 = \left(\sum_{i,j=1}^{n} a_i b_j \beta_{|i-j|} \right)$$

$$= \underbrace{\left(\sum_{k=1}^{n} \sum_{j=1}^{n-k} a_{j+k} b_j \beta_k \right)}_{S_{2,1}} + \underbrace{\left(\sum_{k=1}^{n} \sum_{j=k}^{n} a_{j-k} b_j \beta_k \right)}_{S_{2,2}}. \tag{7}$$

We express $S_{2,1}$ and $S_{2,2}$ as matrix vector products

$$S_{2,1} = \begin{bmatrix} a_2 & a_3 & \cdots & a_{n-1} & a_n & 0 \\ a_3 & a_4 & \cdots & a_n & 0 & 0 \\ \vdots & & & & \vdots \\ a_n & & & & 0 & 0 \\ 0 & & & & 0 & 0 \end{bmatrix} \cdot \begin{bmatrix} b_1 \\ \vdots \\ b_n \end{bmatrix}, \tag{8}$$

$$S_{2,2} = \begin{bmatrix} 0 & a_1 & a_2 & \cdots & a_{n-1} \\ 0 & 0 & a_1 & \cdots & a_{n-2} \\ \vdots & & & & \vdots \\ 0 & 0 & & & a_1 \\ 0 & & & & 0 \end{bmatrix} \cdot \begin{bmatrix} b_1 \\ \vdots \\ b_n \end{bmatrix}. \tag{9}$$

So now we have each of S_1 and S_2 in the required form. We can add $S_{1,1}$ to $S_{2,2}$ and $S_{1,2}$ to $S_{2,1}$ to get the following expression of $S_1 + S_2 = A \times B$.

$$A \times B = (S_{1,1} + S_{2,2}) + (S_{1,2} + S_{2,1})$$

$$= \left(\begin{bmatrix} a_n & a_{n-1} + a_1 & \cdots & a_2 + a_{n-2} & a_1 + a_{n-1} \\ a_1 & a_n & \cdots & a_3 + a_{n-3} & a_2 + a_{n-2} \\ \vdots & & & & \vdots \\ a_{n-2} & \cdots & \cdots & a_n & a_{n-1} + a_1 \\ a_{n-1} & \cdots & \cdots & a_1 & a_n \end{bmatrix} \right.$$

$$\left. + \begin{bmatrix} a_2 & a_3 & \cdots & a_{n-1} & 0 & a_{n-1} \\ a_3 & a_4 & \cdots & 0 & a_{n-1} & a_{n-2} \\ \vdots & & & & \vdots & \\ 0 & a_{n-1} & & & a_2 & a_1 \\ 0 & & & & 0 & 0 \end{bmatrix} \right) \cdot \begin{bmatrix} b_1 \\ \vdots \\ b_n \end{bmatrix}$$

This ends the proof.

Example 1. Let us consider the field $\mathbb{F}_{2^9}$. It is defined as $\mathbb{F}_{2^9} = \mathbb{F}_2[X]/(\beta_9 + 1)$. The Dickson basis of $\mathbb{F}_{2^9}$ is $\mathcal{B} = \{\beta_1, \ldots, \beta_9\}$. The multiplication of two elements A and B can be computed as a matrix vector product. As stated in Theorem 3 the matrix can be decomposed as the sum of a Toeplitz T_A matrix and an essentially Hankel matrix H_A. The Toeplitz matrix T_A is

$$T_A = \begin{bmatrix} a_9 & a_8 + a_1 & a_7 + a_2 & a_6 + a_3 & a_4 + a_5 & a_5 + a_4 & a_6 + a_3 & a_2 + a_7 & a_1 + a_8 \\ a_1 & a_9 & a_8 + a_1 & a_7 + a_2 & a_6 + a_3 & a_4 + a_5 & a_5 + a_4 & a_6 + a_3 & a_2 + a_7 \\ a_2 & a_1 & a_9 & a_8 + a_1 & a_7 + a_2 & a_6 + a_3 & a_4 + a_5 & a_5 + a_4 & a_6 + a_3 \\ a_3 & a_2 & a_1 & a_9 & a_8 + a_1 & a_7 + a_2 & a_6 + a_3 & a_4 + a_5 & a_5 + a_4 \\ a_4 & a_3 & a_2 & a_1 & a_9 & a_8 + a_1 & a_7 + a_2 & a_6 + a_3 & a_4 + a_5 \\ a_5 & a_4 & a_3 & a_2 & a_1 & a_9 & a_8 + a_1 & a_7 + a_2 & a_6 + a_3 \\ a_6 & a_5 & a_4 & a_3 & a_2 & a_1 & a_9 & a_8 + a_1 & a_7 + a_2 \\ a_7 & a_6 & a_5 & a_4 & a_3 & a_2 & a_1 & a_9 & a_8 + a_1 \\ a_8 & a_7 & a_6 & a_5 & a_4 & a_3 & a_2 & a_1 & a_9 \end{bmatrix}$$

and the essentially Hankel matrix H_A is

$$H_A = \begin{bmatrix} a_2 & a_3 & a_4 & a_5 & a_6 & a_7 & a_8 & 0 & a_8 \\ a_3 & a_4 & a_5 & a_6 & a_7 & a_8 & 0 & a_8 & a_7 \\ a_4 & a_5 & a_6 & a_7 & a_8 & 0 & a_8 & a_7 & a_6 \\ a_5 & a_6 & a_7 & a_8 & 0 & a_8 & a_7 & a_6 & a_5 \\ a_6 & a_7 & a_8 & 0 & a_8 & a_7 & a_6 & a_5 & a_4 \\ a_7 & a_8 & 0 & a_8 & a_7 & a_6 & a_5 & a_4 & a_3 \\ a_8 & 0 & a_8 & a_7 & a_6 & a_5 & a_4 & a_3 & a_2 \\ 0 & a_8 & a_7 & a_6 & a_5 & a_4 & a_3 & a_2 & a_1 \\ 0 & 0 & 0 & 0 & 0 & 0 & 0 & 0 & 0 \end{bmatrix}$$

4.2 Dickson Trinomials

Now we assume that the field $\mathbb{F}_{2^n}$ is defined by a three-term irreducible Dickson trinomial P

$$P = 1 + \beta_k + \beta_n, \text{ with } k \leq n/2.$$

In Appendix A we give a list of irreducible trinomials with degree between 163 and 300.

The elements in $\mathbb{F}_{2^n} = \mathbb{F}_2[X]/(P)$ are expressed in the Dickson basis $\mathcal{B} = \{\beta_1, \ldots, \beta_n\}$. Our aim is to express the product of two elements A, and B of $\mathbb{F}_{2^n}$ as Toeplitz or Hankel matrix vector product. We first have

$$C = AB = \underbrace{\left(\sum_{i,j=1}^{n} a_i b_j \beta_{i+j} \right)}_{S_1} + \underbrace{\left(\sum_{i,j=1}^{n} a_i b_j \beta_{|i-j|} \right)}_{S_2}$$

Similar to the previous subsection, here we express S_1 and S_2 as matrix vector product separately. Specifically

1. The sum S_1 is expressed as $Z_A \cdot B$ where Z_A is

$$Z_A = \begin{bmatrix} 0 & 0 & \cdots & 0 & 0 \\ a_1 & 0 & \cdots & 0 & 0 \\ & \vdots & & & \vdots \\ a_{n-1} & \cdots & \cdots & a_1 & 0 \\ a_n & \cdots & \cdots & a_2 & a_1 \\ 0 & a_n & \cdots & a_3 & a_1 \\ & \vdots & & & \vdots \\ 0 & 0 & \cdots & 0 & a_n \end{bmatrix} \begin{matrix} \leftarrow \beta_1 \\ \leftarrow \beta_2 \\ \vdots \\ \leftarrow \beta_n \\ \leftarrow \beta_{n+1} \\ \leftarrow \beta_{n+2} \\ \vdots \\ \leftarrow \beta_{2n} \end{matrix}$$

2. For S_2 we get the same expression as (7)

$$S_2 = \underbrace{\left(\sum_{k=1}^{n} \sum_{j=1}^{n-k} a_{j+k} b_j \beta_k \right)}_{S_{2,1}} + \underbrace{\left(\sum_{k=1}^{n} \sum_{j=k}^{n} a_{j-k} b_j \beta_k \right)}_{S_{2,2}}. \tag{10}$$

where

$$S_2 = \left(\begin{bmatrix} a_2 & a_3 & \cdots & a_{n-1} & a_n & 0 \\ a_3 & a_4 & \cdots & a_n & 0 & 0 \\ & \vdots & & & & \vdots \\ a_n & & & & 0 & 0 \\ 0 & & & & 0 & 0 \end{bmatrix} + \begin{bmatrix} 0 & a_1 & a_2 & \cdots & a_{n-1} \\ 0 & 0 & a_1 & \cdots & a_{n-2} \\ \vdots & & & & \vdots \\ 0 & 0 & & & a_1 \\ 0 & & & & 0 \\ \vdots & & & & \\ 0 & & & & 0 \end{bmatrix} \right) \cdot \begin{bmatrix} b_1 \\ \vdots \\ b_n \end{bmatrix} \tag{11}$$

Now we replace S_1 and S_2 by their corresponding expressions given above in $AB = S_1 + S_2$. We get

$$
AB = \left(
\begin{bmatrix}
0 & 0 & \cdots & 0 & 0 \\
a_1 & 0 & \cdots & 0 & 0 \\
\vdots & & & & \vdots \\
a_{n-1} & \cdots & \cdots & a_1 & 0 \\
a_n & \cdots & \cdots & a_2 & a_1 \\
0 & a_n & \cdots & a_3 & a_1 \\
\vdots & & & & \vdots \\
0 & 0 & \cdots & 0 & a_n
\end{bmatrix}
+
\begin{bmatrix}
0 & a_1 & a_2 & \cdots & a_{n-1} \\
0 & 0 & a_1 & \cdots & a_{n-2} \\
\vdots & & & & \vdots \\
0 & & & & 0 \\
0 & & & & 0 \\
0 & & & & 0 \\
\vdots & & & & \vdots \\
0 & & & & 0
\end{bmatrix}
\right)
\cdot
\begin{bmatrix}
b_1 \\
\vdots \\
b_n
\end{bmatrix}
$$

$$
+
\begin{bmatrix}
a_2 & a_3 & \cdots & a_{n-1} & a_n & 0 \\
a_3 & a_4 & \cdots & a_n & 0 & 0 \\
\vdots & & & & & \vdots \\
a_n & & & 0 & 0 \\
0 & & & 0 & 0 \\
\vdots & & & & \\
0 & & & 0 & 0
\end{bmatrix}
\cdot
\begin{bmatrix}
b_1 \\
\vdots \\
b_n
\end{bmatrix}
\tag{12}
$$

In (12) the addition of two $2n \times n$ Toeplitz matrices results in one single $2n \times n$ Toeplitz matrix. The latter can be horizontally split in the middle to obtain two $n \times n$ Toeplitz matrices, say T_{up} and T_{down}, which can be then multiplied separately with vector $(b_1, \ldots, b_n)$ with a total cost of two $n \times n$ Toeplitz matrix vector products.

The other $2n \times n$ Hankel matrix in (12) has all zero in the lower n rows, contributing nothing to the cost of the matrix vector multiplication. Thus, the total computational cost of (12) is no more than three $n \times n$ Toeplitz or Hankel matrix-vector products.

Remark 2. Among the above three matrices, two of them are triangular. One can attempt to reduce the cost of matrix vector product by using this triangular structure. For example, in the two way split strategy, we can perform $T \cdot V$ as

$$
T \cdot V = \begin{bmatrix} T_0 & T_1 \\ 0 & T_0 \end{bmatrix} \cdot \begin{bmatrix} V_0 \\ V_1 \end{bmatrix} = \begin{bmatrix} T_0 V_0 + T_1 V_1 \\ T_0 V_1 \end{bmatrix}
$$

Such an approach seems to be interesting since the recursive formula needs less computation than in Table 2. However our analysis shows that asymptotically the gain is negligible and the resulting dominant term remains the same as in Table 3.

The reduction

The resulting expression of C in (12) is an unreduced form of $A \times B$, since it has non zero coefficients c_i on rows $i = n+1, \ldots, 2n$. It must be reduced modulo $P = \beta_n + \beta_k + 1$, to get an expression of C in $\mathcal{B}$. We have

$$\beta_i = \beta_n \beta_{i-n} + \beta_{2n-i}$$
$$= (\beta_k + 1)\beta_{i-n} + \beta_{2n-i}$$
$$= \underbrace{\beta_{i-n+k}}_{(R1)} + \underbrace{\beta_{|i-n-k|}}_{(R2)} + \underbrace{\beta_{i-n}}_{(R3)} + \underbrace{\beta_{2n-i}}_{(R4)}$$

In Figure 1 we give the reduction process obtained by replacing in $C = \sum_{i=1}^{2n} c_i \beta_i$ each β_i for $i > n$ by the expression given above.

The process depicted in Figure 1 must be performed two times to get C expressed in the Dickson basis $\mathcal{B}$, since $k \leq n/2$. The full reducing part requires $8n$ XOR gates and is performed in time $6T_X$.

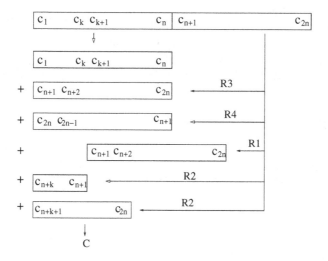

Fig. 1. Dickson Trinomial Reduction Process

Table 4. Complexity of Dickson Multiplier

Method	b	Space		Time
		# AND	# XOR	
Dick. Bin.	2	$2n^{\log_2(3)}$	$11n^{\log_2(3)} - 11n$	$(2\log_2(n) + 1)T_X + T_A$
	3	$2n^{\log_3(6)}$	$48/5 n^{\log_3(6)} - 11n + 3/5$	$(3\log_3(n) + 1)T_X + T_A$
Dick. Tri.	2	$2n^{\log_2(3)}$	$11n^{\log_2(3)} - 4n + 1$	$(2\log_2(n) + 6)T_X + T_A$
	3	$2n^{\log_3(6)}$	$48/5 n^{\log_3(6)} - 2n + 1/5$	$(3\log_3(n) + 6)T_X + T_A$
ONB I [3]	2	$n^{\log_2(3)} + n$	$5.5 n^{\log_2(3)} - 4n - 0.5$	$(2\log_2(n) + 1)T_X + T_A$
	3	$n^{\log_3(6)} + n$	$24/5 n^{\log_3(6)} - 3n - 4/5$	$(3\log_3(n) + 1)T_X + T_A$
ONB II [3]	2	$2n^{\log_2(3)}$	$11n^{\log_2(3)} - 12n + 1$	$(2\log_2(n) + 1)T_X + T_A$
	3	$2n^{\log_3(6)}$	$48/5 n^{\log_3(6)} - 10n - 2/5$	$(3\log_3(n) + 1)T_X + T_A$

5   Complexity and Comparison

In this section we provide the corresponding complexity of each of our multipliers presented in the previous section. The complexities are easily deduced from complexity given in Table 2.

In a recent paper Mullin *et al.* [7] pointed out that there were some links between the Dickson basis and the normal basis. In practice, a Dickson basis is interesting when no optimal normal basis exists for the considered field. This is the case for NIST recommended binary fields $\mathbb{F}_{2^{163}}$ and $\mathbb{F}_{2^{283}}$.

In Table 5 we give fields which can be constructed with a Dickson binomial. In Table 6 we give irreducible Dickson trinomials of low degree. We can remark that NIST fields can be constructed with Dickson trinomials, and thus we obtain a subquadratic multiplier in each of these cases.

We also note that recently a type II optimal normal basis has been presented in [5] using the FFT technique, which normally outperforms other sub-quadratic complexity multipliers for very large values of n. Hardware architectures of *bit-serial* type multipliers using the Dickson basis have been presented in [1].

6   Conclusion

In this paper we have presented new parallel multipliers based on Dickson basis representation of binary fields. The multiplier for an irreducible Dickson binomial has a complexity similar to subquadratic multiplier for ONB II. For an irreducible Dickson trinomial, the multiplier has a slightly more space complexity, but can be used for fields with degree less than 300.

References

1. Ansari, B., Anwar Hasan, M.: Revisiting finite field multiplication using dickson bases. Technical report, University of Waterloo, Ontario, Canada (2007)
2. Dickson, L.E.: The analytic representation of substitutions on a power of a prime number of letters with a discussion of the linear group. Ann. of Math. 11, 161–183 (1883)
3. Fan, H., Hasan, M.A.: A new approach to sub-quadratic space complexity parallel multipliers for extended binary fields. IEEE Trans. Computers 56(2), 224–233 (2007)
4. Fan, H., Hasan, M.A.: Subquadratic computational complexity schemes for extended binary field multiplication using optimal normal bases. In: IEEE Trans. Computers (2008)
5. Giorgi, P., Negre, C., Plantard, T.: Subquadratic binary field multiplier in double polynomial system. In: SECRYPT 2007, Barcelona, Spain (2007)
6. Lild, R., Mullen, G.L., Turnwald, G.: Dickson Polynomials. Pitman Monograpah and Survey n Pure and Applied Mathematic, vol. 65 (1993)
7. Mullin, R.C., Mahalanobis, A.: Dickson bases and finite fields. Technical report, Universite of Waterloo, Ontario (2007)

Appendix

A Binomials and Trinomials for Field Definition

In Table 5 we give the degree $n \in [160, 300]$ of field $\mathbb{F}_{2^n} = \mathbb{F}_2[X]/(P)$ where P satisfies

$$P \times (X + 1) = \beta_{n+1} + 1,$$

β_{n+1} is a Dickson polynomial. For such field, binomial subquadratic multiplier can be used to perform the multiplication.

Table 5. Degrees of fields which admit a binomial subquadratic multiplier

n	167, 173, 198, 196, 190, 198, 238, 252, 262, 268, 270

Table 6. Irreducible Dickson trinomials $\beta_n + \beta_k + 1$

n	k	n	k
163	43, 67, 97, 100, 128, 155	201	84
165	66, 78, 114, 132	202	7,187
167	68, 88	203	5, 107, 113
170	5, 11,25,55,61,71,125,155,157	205	43, 53, 109, 169, 179, 193
171	144	207	18, 180
172	95	208	7,125
173	40,82,85	211	19,85,95
175	26,158	212	73
176	79,89	215	22,64,98,122,166
178	65,73	218	113,127,133,137
179	85	219	120,156
181	35, 115, 134	220	167
183	138	221	14,46,71,145,200,209
184	151	223	82,190
187	28,32,95,115,128,163	224	101
188	73	225	36,72, 144
189	54	226	121,205
191	14,74,106,124,146	227	125,145
193	188	229	50
194	25, 55	231	30,114,156
197	88, 107, 110, 155, 170	235	13,17,32,37,88,103,112,128,173
199	86	237	42
200	7,17,31,77	239	124,164,220

Table 6. (*continued*)

n	k	n	k
241	16,160,176,200	272	7,235,245
242	85, 223	273	240
244	121,169	274	65,101,181,205,269
245	37,43,52,61,116,172,187	275	44,59,88,176,227
247	22,50,110,245	277	70,95,98,118,125,130,175
248	65,137	279	90,234
250	25,85,125,155,175,181,	280	17,103,173,197
250	185,209,217,245	283	37,80,145,155,157,215,95
251	119,145,211	285	42,132
253	7,10, 23, 115, 142, 158, 170, 205	285	246
255	174,186	289	40, 280
256	91,209	290	41, 53, 79, 85 ,113,125,163,185
259	5,20	291	24, 25
259	160	292	133,265
260	97	293	17,55,82,100,140,
261	234	293	227,233,262,275,278
263	20,98,178	295	46,62,154,254
265	112	296	65,221
268	25	298	35,97
269	34,49,125,140,146,190,254	299	119,145
271	46		

Digit-Serial Structures for the Shifted Polynomial Basis Multiplication over Binary Extension Fields

Arash Hariri and Arash Reyhani-Masoleh

Department of Electrical and Computer Engineering
The University of Western Ontario, London, Ontario, Canada
hariri@ieee.org, areyhani@eng.uwo.ca

Abstract. Finite field multiplication is one of the most important operations in the finite field arithmetic. Recently, a variation of the polynomial basis, which is known as the shifted polynomial basis, has been introduced. Current research shows that this new basis provides better performance in designing bit-parallel and subquadratic space complexity multipliers over binary extension fields. In this paper, we study digit-serial multiplication algorithms using the shifted polynomial basis. They include a Most Significant Digit (MSD)-first digit-serial multiplication algorithm and a hybrid digit-serial multiplication algorithm, which includes parallel computations. Then, we explain the hardware architectures of the proposed algorithms and compare them to their existing counterparts. We show that our MSD-first digit-serial shifted polynomial basis multiplier has the same complexity of the Least Significant Digit (LSD)-first polynomial basis multiplier. Also, we present the results for the hybrid digit-serial multiplier which offers almost the half of the latency of the best known digit-serial polynomial basis multipliers.

Keywords: Shifted polynomial basis, multiplication, binary extension fields, digit-serial.

1 Introduction

Finite field arithmetic has an important application in cryptographic algorithms including the elliptic curve cryptography. It has gained lots of interest in the literature, e.g., [1,2,3,4,5,6], and [7]. Designing efficient finite field arithmetic circuits directly affects the performance of the cryptosystems. Multiplication is one of the most important operations in the finite field arithmetic. This operation has been considered by researchers from different points of view. The most common approaches are based on the polynomial basis [3,4,6,8], normal basis [9,10], dual basis [7,11], and the Montgomery multiplication [12,13,14] algorithms. Each of these categories offers different time and area complexities and has its own advantages and disadvantages.

The Shifted Polynomial Basis (SPB) is a variation of the polynomial basis which is proposed in [15]. This basis is constructed by multiplying a polynomial basis by x^{-v}, where x is the root of the irreducible polynomial, v is an integer, and

J. von zur Gathen, J.L. Imaña, and Ç.K. Koç (Eds.): WAIFI 2008, LNCS 5130, pp. 103–116, 2008.

$0 \le v \le m - 1$. In [2], bit-parallel multipliers are designed based on the SPB for irreducible trinomials and type-II pentanomials, which are faster than the best known polynomial basis and dual basis multipliers. Similarly, it is shown in [16] that the squarers designed using the SPB are faster than their polynomial basis counterparts. Using the SPB, a new approach for designing subquadratic area complexity parallel multipliers is outlined in [17], where the reported multipliers are better than the other similar ones in terms of area and time complexities. Also using the SPB, different bit-parallel multipliers are designed for irreducible pentanomials and trinomials in [18] and [19], respectively.

Considering the structure of the algorithm in terms of the number of bits processed at each step, the multipliers over binary extension fields can be classified into three main categories, namely, bit-serial, digit-serial, and bit-parallel multipliers. In bit-serial multipliers, only one bit of the operand is processed in any cycle. This results in reducing the required hardware for implementing the multiplication algorithm. However, bit-serial multipliers are generally slow. Therefore, this type of multiplication algorithms is suitable for the applications where the low-area complexity is preferred over the time complexity. On the other hand, bit-parallel multipliers have opposite properties. In this type of multipliers, the coordinates of the operands are processed in parallel, which results in a good time complexity; however they require much more area than the bit-serial multipliers do.

Digit-serial multipliers are alternatives for bit-serial and bit-parallel multipliers depending on the amount of the resources available. In this type of multipliers, one can trade off between the speed and the area of the multipliers by choosing different digit sizes. In general, greater digit sizes result in faster multipliers with more area; however the hardware overhead is not proportional to the improvement in the time complexity, see for example [20].

In this paper, we study digit-serial shifted polynomial basis multiplication. In this regard, we present the general formulation for the digit-serial multiplication using the shifted polynomial basis and derive an MSD-first digit-serial multiplication algorithm. Then, we choose efficient values to construct the shifted polynomial basis, which reduce the time and area complexities of the general digit-serial multiplication operation. Based on the presented formulation and the algorithm, we also propose an additional digit-serial shifted polynomial basis multiplication algorithm. This multiplication algorithm, which is denoted as hybrid, uses parallel operations to obtain the multiplication product. We compare the proposed multiplication algorithms to the digit-serial polynomial basis multiplication algorithms and show that their complexities match or outperform them. More importantly, the presented hybrid algorithm reduces the latency of the multiplication to half of the latency in polynomial basis multiplication, while it has the same critical path delay.

The rest of this paper is organized as follows. In Section 2, we provide the preliminary background. In Section 3, we consider digit-serial shifted polynomial basis multiplication. Then in Section 4, we provide our discussions and comparisons. Finally, we conclude this paper in Section 5.

2 Preliminaries

In this section, we provide the mathematical formulations to derive our shifted polynomial basis multiplication algorithms. First, we present a short introduction to the binary extension field and then, we explain the shifted polynomial basis.

The binary extension field [21], also known as $GF(2^m)$, is a finite field which includes 2^m field elements. Each $GF(2^m)$ is associated with an irreducible polynomial, i.e., $F(z)$, which is defined over $GF(2)$. The irreducible polynomial $F(z)$ is of degree m and can be represented as

$$F(z) = z^m + f_{m-1}z^{m-1} + \cdots + f_1 z + 1, \tag{1}$$

where $f_i \in \{0, 1\}$ for $i = 1$ to $m - 1$. Assuming x is the root of the irreducible polynomial $F(z)$, i.e., $F(x) = 0$, the set $\{1, x, x^2, \ldots, x^{m-1}\}$ is known as the polynomial basis. This set is used to represent the elements of $GF(2^m)$ as polynomials over $GF(2)$, where $a_i, b_i \in \{0, 1\}$ for $i = 0$ to $m - 1$.

Assuming v is an integer, $0 \le v \le m - 1$, and the set $\{1, x, x^2, \ldots, x^{m-1}\}$ is a polynomial basis for $GF(2^m)$, the Shifted Polynomial Basis (SPB) for $GF(2^m)$ is defined as the set $\{x^{-v}, x^{-v+1}, \ldots, x^{m-v-1}\}$ [15]. Similar to the polynomial basis, it is possible to represent each field element using the SPB. For example, if A and B are two elements of $GF(2^m)$, one can write

$$A = \sum_{i=0}^{m-1} a_i x^{i-v}, B = \sum_{i=0}^{m-1} b_i x^{i-v}, \tag{2}$$

where $a_i, b_i \in \{0, 1\}$ for $i = 0$ to $m - 1$.

The addition of two field elements, represented in the SPB, is carried out by the XOR operation. However, the multiplication of two field elements is complicated and requires more resources. The multiplication in the SPB is defined as $C = \sum_{i=0}^{m-1} c_i x^{i-v} = A \cdot B \bmod F(x)$. The multiplication result, C, is also a field element of degree $m - 1$. This operation can be carried out by using different multiplication algorithms including bit-serial, bit-parallel, and digit-serial algorithms. In the next section, we present digit-serial algorithms for this multiplication.

3 Digit-Serial Shifted Polynomial Basis Multiplication

In a digit-serial multiplier, the bits are grouped as digits and at each cycle, one digit is processed. We define $D \ge 2$ to be the digit size, which means each digit has D bits. We start from the Least Significant Bit (LSB) of the operand B, i.e., b_0, and group D consecutive bits as a digit. This results in having $n = \lceil m/D \rceil$ digits in operand B. Consequently, we obtain

$$B = \sum_{i=0}^{n-1} B_i x^{iD-v}, \tag{3}$$

where

$$B_i = \begin{cases} \sum_{j=0}^{D-1} b_{Di+j}x^j, & 0 \le i \le n-2 \\ \sum_{j=0}^{m-1-D(n-1)} b_{Di+j}x^j, & i = n-1 \end{cases} \tag{4}$$

Using (3) and

$$C = A \cdot B \bmod F(x), \tag{5}$$

one can write the general formulation of the digit-serial SPB multiplication as

$$C = B_0 A x^{-v} + B_1 A x^{D-v} + \cdots + B_{n-1} A x^{(n-1)D-v} \bmod F(x). \tag{6}$$

Now, we try to find appropriate values for v to design efficient digit-serial SPB multipliers. By inspecting (6), we propose to choose $v = (n-1)D$. The reason is that in this case we have $Ax^{(n-1)D-v} = A$ and there is no need to compute $Ax^{(n-1)D-v}$ before processing the digits of B. As a result, we are interested in $(n-1)D - v = 0$, which results in the proposed value for v. Now using $v = (n-1)D$, we rewrite (6) as

$$C = B_0 A x^{-(n-1)D} + B_1 A x^{-(n-2)D} + \cdots + B_{n-1} A \bmod F(x). \tag{7}$$

Now, we can propose a digit-serial SPB multiplication algorithm, namely the Most Significant Digit (MSD)-first digit-serial SPB multiplication algorithm based on (7).

3.1   The MSD-First Digit-Serial SPB Multiplier

Using (7), one can design the MSD-first digit-serial SPB multiplier in which the operand B is processed from its MSD, i.e., B_{n-1}. We show the algorithm corresponding to this multiplier in Algorithm 1 for general irreducible polynomials. Note that $v = (n-1)D$ is chosen to construct the shifted polynomial basis. Step 1 in Algorithm 1 is the initialization and the main operations of the algorithm include a multiplication followed by an addition in Step 3 and a multiplication by x^{-D} followed by a reduction by $F(x)$ in Step 4. In this algorithm, A' and C' can be represented as

$$A' = \sum_{i=0}^{m-1} a_i' x^{i-v}, \quad C' = \sum_{i=0}^{m+D-2} c_i' x^{i-v}. \tag{8}$$

The structure of the MSD-first digit-serial SPB multiplier is shown in Fig. 1. This structure includes two loops. The right and the left loops implement Step 3 and Step 4 of Algorithm 1, respectively. The module represented by $\times$ multiplies A' (a polynomial of degree $m - v - 1$) by a digit of B, i.e., B_i (a polynomial of degree $D-1$), for $i = 0$ to $n-1$, and as a result, its output has $m + D - 1$ bits.

Algorithm 1. The MSD-first digit-serial SPB multiplication

Inputs: A, B, $F(x)$, $n = \lceil m/D \rceil$, $v = (n-1)D$
Output: $C = A \cdot B \bmod F(x)$
Step 1: $A' := A$, $C' := 0$
Step 2: For $i := 0$ to $n-1$
Step 3: $C' := B_{n-i-1}A' + C'$
Step 4: $A' := A' \cdot x^{-D} \bmod F(x)$
Step 5: $C := C' \bmod F(x)$

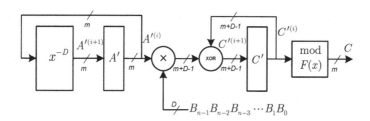

Fig. 1. The MSD-first digit-serial SPB multiplier

This module is shown in Fig. 2a for $m = 11$, $D = 3$, and $v = 9$, where $B_{i,j}$ means the j-th bit of the i-th digit. The module represented by XOR adds the result of the ×-module with the current value of C' and stores it in C' again. In this structure, C' is an $(m+D-1)$-bit register which contains the coordinates of the polynomial C' shown in (8). The x^{-D}-module multiplies A' by x^{-D} and reduces the result by $F(x)$ as shown in Fig. 2b. The final result, shown in (8), is stored in A' using an m-bit register. The final mod $F(x)$ module implements Step 5 of Algorithm 1, which is the final step and is a reduction of a polynomial of degree $(m - v + D - 2)$ by $F(x)$. This operation has a similar structure to Fig. 2b, however in this case, $(D-1)$ terms which are of degree $m - v$ to $(m - v + D - 2)$ should be reduced. As a result, there should be $(D-1)$ rows in Fig. 2b for this operation. Note that in Fig. 1, $A'^{(i)}$ and $C'^{(i)}$ show the content of the registers A' and C' at the i-th iteration of Algorithm 1, respectively.

In Algorithm 1, Step 3 and Step 4 are performed in parallel. As a result, the critical path delay of the multiplier is equal to the maximum of the delays in Step 3 and Step 4. In Step 3 of Algorithm 1, the m-bit A' is multiplied by the D-bit B_{n-i-1} and then, the result is added to C'. Let T_A and T_X represent the delay of a two-input AND gate and the delay of a two-input XOR gate, respectively. This Step requires the delay of T_A to obtain the partial products, and then the delay of $\lceil \log_2(D+1) \rceil T_X$ to sum up D rows of partial products with C' using an XOR tree in the general case (see Fig. 2a). As a result, it requires the delay of $T_A + \lceil \log_2(D+1) \rceil T_X$. In Step 4 of Algorithm 1, the m-bit A' is multiplied by x^{-D} followed by the modulo $F(x)$ reduction. Generally, this can be obtained by the delay of $D(T_A + T_X)$ (see Fig. 2b). Consequently, the multiplier associated

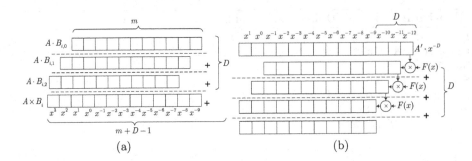

Fig. 2. (a) Multiplication by B_i, (b) multiplication by x^{-D} followed by reduction

with Algorithm 1 has the critical path delay of $D(T_A + T_X)$. Also, the latency of this multiplier is $n + 1$ clock cycles or equivalently, $\lceil m/D \rceil + 1$ clock cycles, including the final mod $F(x)$ operation.

Note that it is possible to do the reduction each time in Step 3 of Algorithm 1, however it increases the critical path delay of the multiplication, especially when the time complexity of the multiplication by x^{-D} followed by the modulo $F(x)$ reduction is optimized.

In Fig. 1, the $\times$-module and the XOR module together require $D \times m$ two-input AND gates and $D \times m$ two-input XOR gates. The x^{-D}-module requires $D \times (m - 1)$ two-input AND gates and $D \times (m - 1)$ two-input XOR gates, and the mod $F(x)$ operation requires $(D - 1) \times (m - 1)$ two-input AND gates and $(D - 1) \times (m - 1)$ two-input XOR gates for the general case of irreducible polynomials. Also, there are $(2m + D - 1)$ registers in this architecture. As a result, we can conclude the following to obtain the complexities of the proposed MSD-first digit-serial SPB multiplier.

Proposition 1. *The MSD-first digit-serial SPB multiplier of Fig. 1 requires $D \times (3m - 2) - m + 1$ two-input AND gates and $D \times (3m - 2) - m + 1$ two-input XOR gates and $(2m + D - 1)$ registers. Also, it has the critical path delay of $D(T_A + T_X)$ and the latency of $n + 1$ clock cycles.*

It is interesting to note that the proposed MSD-first digit-serial SPB multiplication algorithm has the same area and time complexities in comparison with the LSD-first polynomial basis multiplication algorithm proposed in [1].

3.2   Hybrid Digit-Serial SPB Multiplication

From (6), the SPB multiplication can be formulated as

$$C = B_0 A x^{-v} + B_1 A x^{D-v} + \cdots + B_{\lfloor \frac{n}{2} \rfloor} A x^{\lfloor \frac{n}{2} \rfloor D - v} + \cdots + B_{n-1} A x^{(n-1)D - v} \bmod F(x). \tag{9}$$

Now, we choose $v = \lfloor \frac{n}{2} \rfloor D$ to rewrite (9) as

$$
\begin{aligned}
C =& B_0 A x^{-\lfloor \frac{n}{2} \rfloor D} + B_1 A x^{D - \lfloor \frac{n}{2} \rfloor D} + \cdots + B_{\lfloor \frac{n}{2} \rfloor - 1} A x^{-D} + \\
& B_{\lfloor \frac{n}{2} \rfloor} A + B_{\lfloor \frac{n}{2} \rfloor + 1} A x^D \cdots + B_{n-1} A x^{\lfloor \frac{n-2}{2} \rfloor D} \bmod F(x).
\end{aligned}
\tag{10}
$$

It is clear from (10) that C includes two parts. One part is based on the positive powers of x and the other part is based on the negative powers of x. We can show this fact by

$$
C = C' + C'',
\tag{11}
$$

where

$$
\begin{aligned}
C' =& B_0 A x^{-\lfloor \frac{n}{2} \rfloor D} + B_1 A x^{D - \lfloor \frac{n}{2} \rfloor D} + \cdots + \\
& B_{\lfloor \frac{n}{2} \rfloor - 1} A x^{-D} \bmod F(x),
\end{aligned}
\tag{12}
$$

and

$$
\begin{aligned}
C'' =& B_{\lfloor \frac{n}{2} \rfloor} A + B_{\lfloor \frac{n}{2} \rfloor + 1} A x^D + \cdots + \\
& B_{n-1} A x^{\lfloor \frac{n-2}{2} \rfloor D} \bmod F(x).
\end{aligned}
\tag{13}
$$

We note that obtaining C' is a digit-serial SPB multiplication which considers the $\lfloor \frac{n}{2} \rfloor$ least significant digits of the operand B. On the other hand, obtaining C'' is a digit-serial polynomial basis multiplication which involves the $n - \lfloor \frac{n}{2} \rfloor$ most significant digits of the operand B. As explained in the previous section, these two parallel operations can be implemented with an equal critical path delay. A similar approach is outlined in [13] for the digit-serial Montgomery multiplication over binary extension fields. However, two parallel operations of the algorithm in [13] have different critical path delays for general irreducible polynomials. For example, in the simplest case, i.e., one-bit digits, one of the parallel operations (the polynomial basis multiplication) has the critical path delay of $T_A + T_X$, but the other one (the Montgomery multiplication) has the critical path delay of $2T_A + 2T_X$. Also, this technique is applied on the Montgomery multiplication of integers in [22].

Now, based on our proposed MSD-first digit-serial SPB multiplier and also the available LSD-first digit-serial polynomial basis, e.g., [1], we propose an algorithm to reduce the time complexity of the digit-serial SPB multiplication. This algorithm is shown in Algorithm 2. Note that in this algorithm B_{-1} and B_n are equal to zero. It is seen from Algorithm 2 that two multiplications are carried out in parallel and two partial products are summed up and reduced by $F(x)$ in Step 7. In this algorithm, A', A'', C', and C'' can be represented as

$$A' = \sum_{i=0}^{m-1} a_i' x^{i-v}, C' = \sum_{i=0}^{m+D-2} c_i' x^{i-v},$$

$$A'' = \sum_{i=0}^{m-1} a_i'' x^{i-v}, C'' = \sum_{i=0}^{m+D-2} c_i'' x^{i-v}.$$

Algorithm 2. Hybrid digit-serial SPB multiplication

Inputs: A, B, $F(x)$, $n = \lceil m/D \rceil$, $v = \lfloor \frac{n}{2} \rfloor D$
Output: $C = A \cdot B \bmod F(x)$
Step 1: $A' := A$, $C' := 0$, $C'' = 0$, $A'' := A$
Step 2: For $i := 0$ to $\lfloor \frac{n}{2} \rfloor$
Step 3: $A' := A' \cdot x^{-D} \bmod F(x)$
Step 4: $C' := C' + B_{\lfloor \frac{n}{2} \rfloor - 1 - i} A'$
Step 5: $C'' := C'' + B_{\lfloor \frac{n}{2} \rfloor + i} A''$
Step 6: $A'' := A'' \cdot x^{D} \bmod F(x)$
Step 7: $C := C' + C'' \bmod F(x)$

The hardware structure of Algorithm 2 is shown in Fig. 3. In this figure, the top structure obtains (12) and the bottom structure obtains (13). The modules of this figure are similar to the ones used in Fig. 1. The module labeled x^D performs a multiplication by x^D followed by a reduction by $F(x)$. Also, A' and A'' are m-bit registers, whereas C' and C'' are $(M + D - 1)$-bit registers.

It is noted that for odd values of n, there are $\lfloor \frac{n}{2} \rfloor$ terms in (12) and $\lfloor \frac{n}{2} \rfloor + 1$ terms in (13). Obtaining C' requires $\lfloor \frac{n}{2} \rfloor + 1$ clock cycles and it is because the polynomial $Ax^{-D} \bmod F(x)$ should be pre-computed. As a result, a zero is fed to the top structure in Fig. 3 to perform the pre-computation. But, the first term in obtaining C'' is $B_{\lfloor \frac{n}{2} \rfloor} A$ which does not require any pre-computation. Consequently, both C' and C'' can be obtained after $\lfloor \frac{n}{2} \rfloor + 1$ clock cycles. For even values of n, both (12) and (13) include $\frac{n}{2}$ terms. However, C' needs $\frac{n}{2} + 1$ clock cycles and C'' requires $\frac{n}{2}$ clock cycles. We explain the complexity of this algorithm in the next section.

4 Discussion and Comparison

In this section, we consider the time complexity of the proposed digit-serial SPB multipliers in more details and extend the results of [1] to the proposed digit-serial SPB multipliers.

The main operation in Algorithm 1 is the multiplication by x^{-D} followed by a reduction by $F(x)$. Thus, by making this operation faster, one can reduce the

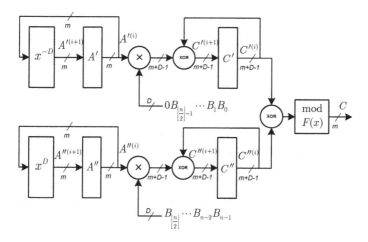

Fig. 3. Architecture of the hybrid digit-serial SPB multiplication (Algorithm 2)

critical path delay of the proposed multipliers. Assuming $T \in GF(2^m)$, we have the following

$$
\begin{aligned}
T \cdot x^{-D} &= (t_{m-1}x^{m-v-1} + \cdots + t_D x^{D-v} + t_{D-1}x^{D-v-1} + \cdots \\
&\quad + t_1 x^{-v+1} + t_0 x^{-v}) \cdot x^{-D} \bmod F(x), \\
&= (t_{m-1}x^{m-v-D-1} + \cdots + t_D x^{-v} + t_{D-1}x^{-v-1} + \cdots \\
&\quad + t_1 x^{-v-D+1} + t_0 x^{-v-D}) \bmod F(x).
\end{aligned} \tag{14}
$$

There are D terms in (14) which should be reduced by $F(x)$ i.e., $(t_{D-1}x^{-v-1} + \cdots + t_1 x^{-v-D+1} + t_0 x^{-v-D}) \bmod F(x)$. As a result, the complexity of (14) depends on the irreducible polynomial $F(x)$ and the value of D. In this regard, we present the following proposition.

Proposition 2. *Assume* $F(z) = z^m + \sum\limits_{i=l+1}^{m-1} f_i z^i + f_l z^l + 1$ *is an irreducible polynomial over* $GF(2)$ *and* x *is a root of* $F(z)$. *In this case, no reduction is required to represent* x^{-v-k} *in the shifted polynomial basis if* $k \le l$.

Proof. We can write

$$
F(x) = x^m + \sum_{i=l+1}^{m-1} f_i x^i + f_l x^l + 1 = 0,
$$

$$
\Rightarrow x^{-v-k} \times (x^m + \sum_{i=l+1}^{m-1} f_i x^i + f_l x^l + 1) = 0,
$$

or

$$x^{m-v-k} + \sum_{i=l+1}^{m-1} f_i x^{i-v-k} + f_l x^{l-v-k} = x^{-v-k}.$$

So, if $l - v - k \geq -v$, then the left side of the equation above is already in the SPB, Thus, no reduction is required for $l \geq k$ and the proof is complete. ∎

Now, we can propose the following lemma which is used to obtain the complexity results of the proposed digit-serial SPB multipliers.

Lemma 1. *Let $\{x^{-v}, x^{-v+1}, \ldots, x^{m-v-1}\}$ be the SPB and A be a field element, where x is a root of the irreducible polynomial $F(z) = z^m + \sum_{i=l+1}^{m-1} f_i z^i + f_l z^l + 1$. Then, $A \cdot x^{-D} \bmod F(x)$ can be represented in the shifted polynomial basis by only one step of reduction if $D \leq l$, where D is the digit size. In this case, $A \cdot x^{-D} \bmod F(x)$ is obtained with the delay of $T_A + \lceil \log_2(D+1) \rceil T_X$ for the general case.*

Proof. We can represent $A \in GF(2^m)$ as

$$A = a_{m-1} x^{m-v-1} + \cdots + a_2 x^{-v+2} + a_1 x^{-v+1} + a_0 x^{-v},$$

and consequently,

$$A \cdot x^{-D} = a_{m-1} x^{m-v-1-D} + \cdots + a_2 x^{-v+2-D} + a_1 x^{-v+1-D} + a_0 x^{-v-D}.$$

By using proposition 2, it is clear that the terms whose powers of x are between $-v-1$ and $-v-D$, i.e., $a_{D-1} x^{-v-1}$ and $a_0 x^{-v-D}$, can be represented in the shifted polynomial basis by only one step of reduction if $D \leq l$. These D terms can be reduced in parallel with the delay of T_A and then, they should be summed up with the other term of $A \cdot x^{-D}$. This requires the total delay of $T_A + \lceil \log_2(D+1) \rceil T_X$. ∎

Obtaining $A \cdot x^{-D} \bmod F(x)$ for $D \leq l$ is depicted in Fig. 4. In this case, D terms should be reduced by the irreducible polynomial which requires $D \times (m-l)$ two-input AND gates as $f_m = 1$. Then, they should be added to the rest of the terms in $A \cdot x^{-D}$ by using $D \times (m-l+1)$ XOR gates.

Remark 1. The area and time complexities of the proposed hybrid digit-serial SPB multiplier can be obtained using the results presented for the MSD-first digit-serial SPB multipliers. This algorithm has the critical path of the MSD-first digit-serial SPB multiplication algorithm, however its latency is almost the half of that of the MSD-first digit-serial SPB multiplication algorithm. One can achieve this latency using the LSD-first digit-serial polynomial basis multipliers if the digit size $2D$ is chosen. However, this results in doubling the critical path delay in the general case or adding an extra delay of an XOR gate in the special

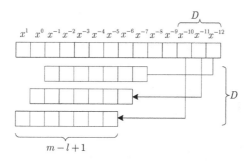

Fig. 4. Multiplication by x^{-D} followed by reduction for $D \leq l$

cases. The hardware overhead of the hybrid digit-serial SPB multiplier with the digit size D in comparison to the digit-serial polynomial basis multiplier with digit size $2D$ is $2m$ registers and $m+D-1$ XOR gates. In general, the time×area factor of the hybrid digit-serial SPB multiplier is equal to that of the MSD-first digit-serial SPB multiplier. However, better results can be achieved if different structures like semi-systolic arrays used to implement the hybrid digit-serial algorithms. This is because the lower latency results in reducing the number of the required rows of the semi-systolic array and as a result, even for equal digit sizes, the hardware overhead will be very low.

By using Lemma 1, the time complexity of the proposed digit-serial SPB multipliers is presented in Table 1.

It is possible to construct the shifted polynomial basis using $v = m - 1$ which extends the range of the efficient digit sizes. In this case, the operand B is

Table 1. Time Complexity of the Digit-Serial Multipliers over Binary Extension Fields

Algorithm	Type	Critical Path delay	Latency
$F(z) = z^m + f_w z^w + \sum\limits_{i=l+1}^{w-1} f_i z^i + f_l z^l + 1, D > \min\{l, m-w\}$			
Algorithm 1: MSD-first	SPB	$D(T_A + T_X)$	$n+1$
Algorithm 2: Hybrid	SPB	$D(T_A + T_X)$	$\lfloor \frac{n}{2} \rfloor + 2$
$F(z) = z^m + f_w z^w + \sum\limits_{i=1}^{w-1} f_i z^i + 1, D > m - w$			
MSD-first [1]	Polynomial basis	$D(T_A + T_X) + T_X$	$n+1$
LSD-first [1]	Polynomial basis	$D(T_A + T_X)$	$n+1$
$F(z) = z^m + f_w z^w + \sum\limits_{i=l+1}^{w-1} f_i z^i + f_l z^l + 1, 2 \leq D \leq \min\{l, m-w\}$			
Algorithm 1: MSD-first	SPB	$T_A + \lceil \log_2(D+1) \rceil T_X$	$n+1$
Algorithm 2: Hybrid	SPB	$T_A + \lceil \log_2(D+1) \rceil T_X$	$\lfloor \frac{n}{2} \rfloor + 2$
$F(z) = z^m + f_w z^w + \sum\limits_{i=1}^{w-1} f_i z^i + 1, 2 \leq D \leq m - w$			
MSD-first [1]	Polynomials basis	$T_A + \lceil \log_2(2D+1) \rceil T_X$	$n+1$
LSD-first [1]	Polynomials basis	$T_A + \lceil \log_2(D+1) \rceil T_X$	$n+1$

represented as $B = b_{m-1} + b_{m-2}x^{-1} + \cdots + a_0 x^{-(m-1)}$. This time, instead of grouping the bits from right to left (e.g., starting from the LSB), we start from the MSB of B and group D consecutive terms to form a digit of degree at most $-(D-1)$, i.e.,

$$
B_i' = \begin{cases}
\sum_{j=0}^{D-1} b_{m-Di-j-1}x^{-j}, & 0 \leq i \leq n-2 \\
\sum_{j=0}^{m-1-D(n-1)} b_{m-Di-j-1}x^{-j}, & i = n-1
\end{cases}
\tag{15}
$$

So, $B = B_0' + B_1'x^{-D} + .. + B_{n-1}'x^{-(n-1)D}$. An algorithm similar to Algorithm 1 can be used as well. However in this case, the coefficients of C' in Step 3 of Algorithm 1 have degrees between $-v - D + 1$ and $m - v - D$. This is depicted in Fig. 5a for $m = 11$, $v = 10$, and $D = 5$, where $B_{i,j}'$ represents the j-th bit of B_i'. Note that the partial products are shifted to the right in this case. The complexity of the multiplication of a field element by B_i' is the same as the one shown in Fig. 2a. Therefore, the reductions in Steps 3 and 4 of Algorithm 1 are similar in this case and as a result, the digit size should satisfy $2 \leq D \leq l$ for the fast multiplication.

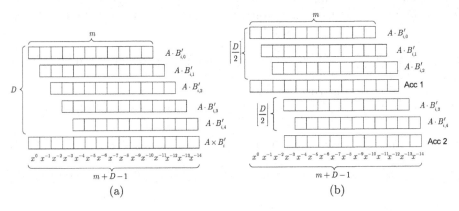

Fig. 5. Multiplication by B_i' using (a) single accumulator, (b) double accumulator

An example: We use $m = 163$ which is recommended by NIST for elliptic curve digital signatures algorithm [23]. Considering $F(z) = z^{163} + z^{97} + z^{96} + z^{95} + 1$ as an irreducible pentanomial, the digit size in Algorithm 1 should satisfy $2 \leq D \leq 66$ which results in efficient implementation. For the digit-serial polynomial basis multiplication algorithms of [1], the digit size should satisfy $2 \leq D \leq 66$ to provide the same complexity. As a result, the digit-size for the hybrid digit-serial SPB multipliers should satisfy $2 \leq D \leq 66$. Using $v = m - 1$ and grouping the coordinates of B from left to right, the digit size should satisfy $2 \leq D \leq 95$.

The techniques introduced in [8] can be extended to the SPB to reduce the time complexity of the digit-serial multipliers as well. In this case, multiple accumulators are used to implement the multiplication $A \times B_i'$. This is shown in Fig. 5b using two accumulators. The main difference is that the results of [8] are presented for the multiplication by x^D followed by a reduction. However, it is possible to extend them to the multiplication by x^{-D} followed by a reduction used in the SPB.

5   Conclusions

In this paper, we have studied the SPB multiplication over binary extension fields and proposed digit-serial multiplication algorithms. In this regard, we have proposed two digit-serial SPB multiplication algorithms. The proposed MSD-first digit-serial algorithm is as efficient as the LSD-first polynomial basis multiplication algorithm, which is the fastest algorithm for digit-serial polynomial multiplication. Also, we have studied the possible cases to reduce the complexity of the digit-serial SPB multipliers based on the chosen digit size and the irreducible polynomial.

We have also proposed a hybrid algorithm which uses parallel computations to make the multiplication process faster. This algorithm has half of the latency of the LSD-first digit-serial polynomial basis multiplier with the same critical path delay, as one of the fastest digit-serial polynomial basis multipliers.

Acknowledgment

The authors would like to thank the reviewers for their constructive comments. This work has been supported in part by an NSERC Discovery grant awarded to Arash Reyhani-Masoleh.

References

1. Song, L., Parhi, K.: Low-Energy Digit-Serial/Parallel Finite Field Multipliers. The Journal of VLSI Signal Processing 19(2), 149–166 (1998)
2. Fan, H., Hasan, M.: Fast Bit Parallel Shifted Polynomial Basis Multipliers in $GF(2^n)$. IEEE Transactions on Circuits and Systems I: Fundamental Theory and Applications 53(12), 2606–2615 (2006)
3. Imana, J., Sanchez, J.: Bit-Parallel Finite Field Multipliers for Irreducible Trinomials. IEEE Transactions on Computers 55(5), 520–533 (2006)
4. Reyhani-Masoleh, A., Hasan, M.: Low Complexity Bit Parallel Architectures for Polynomial Basis Multiplication over $GF(2^m)$. IEEE Transactions on Computers 53(8), 945–959 (2004)
5. Yeh, C.S., Reed, I.S., Truong, T.K.: Systolic Multiplier for Finite Fields $GF(2^m)$. IEEE Transactions on Computers C-33, 357–360 (1983)
6. Beth, T., Gollman, D.: Algorithm Engineering for Public Key Algorithms. IEEE Journal on Selected Areas in Communications 7(4), 458–466 (1989)
7. Rodriguez-Henriguez, F., Koc, C.: Parallel Multipliers Based on Special Irreducible Pentanomials. IEEE Transactions on Computers 52(12), 1535–1542 (2003)

8. Kumar, S., Wollinger, T., Paar, C.: Optimum Digit Serial $GF(2^m)$ Multipliers for Curve-Based Cryptography. IEEE Transactions on Computers 55(10), 1306–1311 (2006)

9. Koc, C.K., Sunar, B.: Low-Complexity Bit-Parallel Canonical and Normal Basis Multipliers for a Class of Finite Fields. IEEE Transactions on Computers 47(3), 353–356 (1998)

10. Sunar, B., Koc, C.K.: An Efficient Optimal Normal Basis Type II Multiplier. IEEE Transactions on Computers 50(1), 83–87 (2001)

11. Wu, H., Hasan, M., Blake, I.: New Low-Complexity Bit-Parallel Finite Field Multipliers Using Weakly Dual Bases. IEEE Transactions on Computers 47(11), 1223–1234 (1998)

12. Koc, C., Acar, T.: Montgomery Multiplication in $GF(2^k)$. Designs, Codes and Cryptography 14(1), 57–69 (1998)

13. Batina, L., Mentens, N., Preneel, B., Verbauwhede, I.: Balanced Point Operations for Side-Channel Protection of Elliptic Curve Cryptography. Information Security, IEE Proceedings 152(1), 57–65 (2005)

14. Horng, J.S., Lu, E.H.: Low-Complexity Bit-Parallel Systolic Montgomery Multipliers for Special Classes of $GF(2^m)$. IEEE Transactions on Computers 54(9), 1061–1070 (2005)

15. Fan, H., Dai, Y.: Fast Bit-Parallel $GF(2^n)$ Multiplier for All Trinomials. IEEE Transactions on Computers 54(4), 485–490 (2005)

16. Park, S., Chang, K.: Low Complexity Bit-Parallel Squarer for $GF(2^n)$ Defined by Irreducible Trinomials. IEICE Transactions on Fundamentals of Electronics, Communications and Computer Sciences 89, 2451–2452 (2006)

17. Fan, H., Hasan, M.: A New Approach to Subquadratic Space Complexity Parallel Multipliers for Extended Binary Fields. IEEE Transactions on Computers 56(2), 224–233 (2007)

18. Park, S., Chang, K., Hong, D.: Efficient Bit-Parallel Multiplier for Irreducible Pentanomials Using a Shifted Polynomial Basis. IEEE Transactions on Computers 55(9), 1211–1215 (2006)

19. Negre, C.: Efficient Parallel Multiplier in Shifted Polynomial Basis. Journal of Systems Architecture 53(2-3), 109–116 (2007)

20. Sakiyama, K., Batina, L., Mentens, N., Preneel, B., Verbauwhede, I.: Small-Footprint ALU for Public-Key Processors for Pervasive Security. In: Workshop on RFID Security, pp. 77–88 (2006)

21. Lidl, R., Niederreiter, H.: Introduction to Finite Fields and Their Applications. Cambridge University Press, New York (1986)

22. Kaihara, M.E., Takagi, N.: Bipartite Modular Multiplication Method. IEEE Transactions on Computers 57(2), 157–164 (2008)

23. Recommended Elliptic Curves for Federal Government Use, csrc.nist.gov/encryption/dss/ecdsa/NISTReCur.pdf

Some Theorems on Planar Mappings

Gohar M. Kyureghyan and Alexander Pott

Department of Mathematics, Otto-von-Guericke University of Magdeburg,
Universitätplatz 2, 39106 Magdeburg, Germany
{gohar.kyureghyan,alexander.pott}@ovgu.de

Abstract. A mapping $f : \mathbb{F}_p^n \rightarrow \mathbb{F}_p^n$ is called planar if for every nonzero $a \in \mathbb{F}_p^n$ the difference mapping $D_{f,a} : x \mapsto f(x + a) - f(x)$ is a permutation of $\mathbb{F}_p^n$. In this note we prove that two planar functions are CCZ-equivalent exactly when they are EA-equivalent. We give a sharp lower bound on the size of the image set of a planar function. Further we observe that all currently known main examples of planar functions have image sets of that minimal size.

Keywords: Planar mapping, Perfect nonlinear mapping, CCZ-equivalence, Image set.

1 Introduction

Let p be an odd prime. Given a mapping $f : \mathbb{F}_p^n \rightarrow \mathbb{F}_p^n$ and a nonzero element $a \in \mathbb{F}_p^n$, we call the mapping

$$D_{f,a} : \mathbb{F}_p^n \rightarrow \mathbb{F}_p^n, \quad x \mapsto f(x + a) - f(x)$$

the difference mapping of f defined by a. A mapping $f : \mathbb{F}_p^n \rightarrow \mathbb{F}_p^n$ is called *planar* if all its difference mappings are bijective. Planar mappings were introduced in [6] to describe projective planes with certain properties.

Two mappings $f, g : \mathbb{F}_p^n \rightarrow \mathbb{F}_p^n$ are called *extended affine equivalent* (EA-equivalent), if $g = A_1 \circ f \circ A_2 + A$ for some affine permutations A_1, A_2 and an affine mapping A. It is easy to see that if f is a planar mapping then all functions EA-equivalent to it are planar as well.

Currently known EA-inequivalent planar polynomials over finite fields are:

(a) x^2 in $\mathbb{F}_{p^n}$ (folklore)

(b) x^{p^k+1} in $\mathbb{F}_{p^n}$, $k \leq n/2$ and $n/(k,n)$ is odd ([6],[5])

(c) $x^{10} + x^6 - x^2$ in $\mathbb{F}_{3^n}$, $n \geq 5$ is odd ([5])

(d) $x^{10} - x^6 - x^2$ in $\mathbb{F}_{3^n}$, $n \geq 5$ is odd ([9], [3])

(e) $x^{p^s+1} - u^{p^k-1}x^{p^k+p^{2k+s}}$ in $\mathbb{F}_{p^{3k}}$, where $(k,3) = 1$, $k - s \equiv 0$ (mod 3) and $k/(k,s)$ is odd, and u is a primitive element of $\mathbb{F}_{p^{3k}}$ ([14])

(f) $x^{(3^k+1)/2}$ in $\mathbb{F}_{3^n}$, $k \geq 3$ is odd and $(k,n) = 1$ ([5],[10]).

J. von zur Gathen, J.L. Imaña, and Ç.K. Koç (Eds.): WAIFI 2008, LNCS 5130, pp. 117–122, 2008.

Note that the polynomials in (a)-(e) are of shape

$$\sum_{i,j=0}^{n-1} a_{i,j} x^{p^i + p^j}, \ a_{i,j} \in \mathbb{F}_{p^n}.$$

The polynomials of this type are called Dembowski-Ostrom polynomials. In [3] it is proved that the classification of planar Dembowski-Ostrom polynomials is equivalent to the classification of finite commutative semifields of odd order. A planar mapping of $\mathbb{F}_p^n$ with the image set of size $(p^n + 1)/2$ yields either a skew Hadamard difference set or a Paley type partial difference set depending on p^n (mod 4) as shown in [9],[13]. Planar mappings are used to construct optimal constant-composition codes and signal sets in [8],[7].

2   EA- and CCZ-Equivalence

If $h : \mathbb{F}_2^n \to \mathbb{F}_2^n$ then the image set of a difference mapping $D_{h,a}$ has at most 2^{n-1} elements. Indeed, note that $D_{h,a}(x) = D_{h,a}(x + a)$ for any $x \in \mathbb{F}_2^n$. A mapping $h : \mathbb{F}_2^n \to \mathbb{F}_2^n$ is called almost perfect nonlinear (APN) if the image set of every difference mapping $D_{h,a}$ is of maximal size 2^{n-1}. APN mappings are of interest in cryptology since they provide the optimal resistance against differential attacks. It is easy to see that the APN property is invariant under EA-equivalence. In [2] an extension of EA-equivalence was introduced:

Two mappings $f, g : \mathbb{F}_p^n \to \mathbb{F}_p^n$ are called Carlet-Charpin-Zinoviev equivalent (CCZ-equivalent) if the set $\left\{ \begin{pmatrix} x \\ g(x) \end{pmatrix} \mid x \in \mathbb{F}_p^n \right\} \subset \mathbb{F}_p^{2n}$ is the image of the set $\left\{ \begin{pmatrix} x \\ f(x) \end{pmatrix} \mid x \in \mathbb{F}_p^n \right\} \subset \mathbb{F}_p^{2n}$ under an affine permutation of $\mathbb{F}_p^{2n}$. In other words, two mappings of $\mathbb{F}_p^n$ are CCZ-equivalent if their graphs are affine equivalent in $\mathbb{F}_p^{2n}$.

In [2] it is shown that CCZ-equivalent mappings have the same differential properties and that EA-equivalence is a special case of CCZ-equivalence. In [1] it is shown that CCZ-equivalence does not coincide with EA-equivalence for APN mappings. The authors of this note have been asked frequently whether CCZ-equivalence allows to construct EA-inequivalent mappings from a known planar mapping. The answer is negative as it is shown below using arguments from [1].

Given a mapping $f : \mathbb{F}_p^n \to \mathbb{F}_p^n$ and a subset $S \subseteq \mathbb{F}_p^n$, we denote by $f(S)$ the image set of S.

Lemma 1. *Let* $f, g : \mathbb{F}_p^n \to \mathbb{F}_p^n$ *be planar mappings. Set*

$$F = \left\{ \begin{pmatrix} x \\ f(x) \end{pmatrix} \mid x \in \mathbb{F}_p^n \right\}, \quad G = \left\{ \begin{pmatrix} x \\ g(x) \end{pmatrix} \mid x \in \mathbb{F}_p^n \right\}$$

and

$$O = \left\{ \begin{pmatrix} 0 \\ y \end{pmatrix} \mid y \in \mathbb{F}_p^n \right\}.$$

If $\mathcal{L} : \mathbb{F}_p^{2n} \to \mathbb{F}_p^{2n}$ *is a linear permutation satisfying* $\mathcal{L}(F) = G$ *then* $\mathcal{L}(O) = O$.

Proof. Let a be a fixed nonzero element of $\mathbb{F}_p^n$. Then for any $b \in \mathbb{F}_p^n$ there is $x \in \mathbb{F}_p^n$ such that $b = f(x + a) - f(x)$. Thus it holds

$$\mathcal{L}\begin{pmatrix} a \\ b \end{pmatrix} = \mathcal{L}\begin{pmatrix} x + a - x \\ f(x + a) - f(x) \end{pmatrix} = \mathcal{L}\begin{pmatrix} x + a \\ f(x + a) \end{pmatrix} - \mathcal{L}\begin{pmatrix} x \\ f(x) \end{pmatrix}$$

$$= \begin{pmatrix} y \\ g(y) \end{pmatrix} - \begin{pmatrix} y' \\ g(y') \end{pmatrix} \notin O,$$

where we used that $y \neq y'$ since $\mathcal{L}\begin{pmatrix} x + a \\ f(x + a) \end{pmatrix} \neq \mathcal{L}\begin{pmatrix} x \\ f(x) \end{pmatrix}$. The assumption that $\mathcal{L}$ is bijective implies the statement. □

Theorem 1. *Let $f, g : \mathbb{F}_p^n \to \mathbb{F}_p^n$ be planar mappings. If f and g are CCZ-equivalent then they are EA-equivalent.*

Proof. Let $\mathcal{A} : \mathbb{F}_p^{2n} \to \mathbb{F}_p^{2n}$ be an affine mapping which maps the graph of f onto the graph of g. Then $\mathcal{A}\begin{pmatrix} x \\ y \end{pmatrix} = \mathcal{L}\begin{pmatrix} x \\ y \end{pmatrix} + \begin{pmatrix} c_1 \\ c_2 \end{pmatrix}$, where $\mathcal{L} : \mathbb{F}_p^{2n} \to \mathbb{F}_p^{2n}$ is linear and $\begin{pmatrix} c_1 \\ c_2 \end{pmatrix}$ is a fixed element of $\mathbb{F}_p^{2n}$. Note that

$$\mathcal{L}\begin{pmatrix} x \\ f(x) \end{pmatrix} = \begin{pmatrix} y - c_1 \\ g(y) - c_2 \end{pmatrix} = \begin{pmatrix} y' \\ g(y' + c_1) - c_2 \end{pmatrix}.$$

Thus $\mathcal{L}$ maps the graph of f onto the graph of $g(y' + c_1) - c_2$ which is EA-equivalent to g and hence is planar. Let a matrix representation of $\mathcal{L}$ be given by

$$\begin{pmatrix} L_1 & L_2 \\ L_3 & L_4 \end{pmatrix},$$

where L_i is an $n \times n$ matrix over $\mathbb{F}_p$. Then by Lemma 1 the matrix L_2 must be the zero matrix. Hence

$$\mathcal{A}\begin{pmatrix} x \\ f(x) \end{pmatrix} = \begin{pmatrix} L_1 & 0 \\ L_3 & L_4 \end{pmatrix}\begin{pmatrix} x \\ f(x) \end{pmatrix} + \begin{pmatrix} c_1 \\ c_2 \end{pmatrix}$$

$$= \begin{pmatrix} L_1(x) + c_1 \\ L_3(x) + L_4(f(x)) + c_2 \end{pmatrix} = \begin{pmatrix} y \\ g(y) \end{pmatrix}.$$

This shows that $g(L_1(x) + c_1) = L_3(x) + L_4(f(x)) + c_2$ with L_1 and L_4 of full rank, completing the proof. □

3   On the Image Set of a Planar Mapping

Let p be an odd prime. Given a set $S \subseteq \mathbb{F}_p^n$ and $a \in \mathbb{F}_p^n$, we denote by $f^{-1}(S)$ the preimage of S and by $w(a)$ the size of $f^{-1}(\{a\})$. A mapping $f : \mathbb{F}_p^n \to \mathbb{F}_p^n$ is called 2-to-1 if there is a unique a_0 with $w(a_0) = 1$ and $w(a) \in \{0, 2\}$ for the remaining a.

In this section we observe that results and techniques from [13] imply that all currently known planar mappings are 2-to-1 up to addition of a linear mapping.

Lemma 2 ([13]). *Let $f : \mathbb{F}_p^n \to \mathbb{F}_p^n$ be a planar mapping. Then*

$$\sum_{x \in \mathbb{F}_p^n} w(x+b)w(x) = \begin{cases} p^n - 1 & \text{if } b \neq 0 \\ 2p^n - 1 & \text{if } b = 0. \end{cases}$$

Proof. The following is a slightly simplified version of the proof from [13]. Note that

$$\sum_{x \in \mathbb{F}_p^n} w(x+b)w(x) = \sum_{x \in \mathbb{F}_p^n} |\{(u,v) \in \mathbb{F}_p^n \times \mathbb{F}_p^n \mid f(u) = x + b, \ f(v) = x\}|$$

$$= |\{(u,v) \in \mathbb{F}_p^n \times \mathbb{F}_p^n \mid f(u) - f(v) = b\}|$$

$$= |\{(a,v) \in \mathbb{F}_p^n \times \mathbb{F}_p^n \mid f(v+a) - f(v) = b\}|.$$

Then the statement follows from the observation that for any fixed $a \neq 0$ there is exactly one $v \in \mathbb{F}_p^n$ satisfying $f(v+a) - f(v) = b$. If $b = 0$, then $(0,v), v \in \mathbb{F}_p^n$, satisfy the equation as well. □

The next result extends Lemma 2.3 from [13].

Theorem 2. *Let $f : \mathbb{F}_p^n \to \mathbb{F}_p^n$ be a planar mapping and I be its image set. Then*

$$|I| \geq \frac{p^n + 1}{2}.$$

Moreover, $|I| = \frac{p^n+1}{2}$ if and only if f is 2-to-1.

Proof. By Lemma 2

$$\sum_{y \in \mathbb{F}_p^n} w^2(y) = \sum_{y \in I} w^2(y) = 2p^n - 1. \tag{1}$$

Obviously it holds also

$$\sum_{y \in I} w(y) = p^n. \tag{2}$$

Then (1) and (2) imply

$$0 \leq \sum_{y \in I} (w(y) - 2)^2 = 2p^n - 1 - 4p^n + 4|I| = 4|I| - 2p^n - 1, \tag{3}$$

and thus

$$|I| \geq \left\lceil \frac{2p^n - 1}{4} \right\rceil = \frac{p^n + 1}{2},$$

proving the first statement. Suppose now that $|I| = \frac{p^n+1}{2}$. Then (3) is reduced to

$$\sum_{y \in I} (w(y) - 2)^2 = 4|I| - 2p^n - 1 = 4\frac{p^n + 1}{2} - 2p^n - 1 = 1.$$

The equality $\sum_{y \in I}(w(y) - 2)^2 = 1$ is possible if and only if there is a unique $y_0 \in I$ with $(w(y_0) - 2)^2 = 1$ and $(w(y) - 2)^2 = 0$ (and hence $w(y) = 2$) for the remaining elements of I. Then from (2) it follows that $w(y_0) = 1$, proving the second part of the statement. □

Corollary 1. *Let $f(x) = \sum_{i=0}^{p^n-1} a_i x^i$, $a_i \in \mathbb{F}_{p^n}$, be a planar polynomial. Suppose $a_i = 0$ for all odd i. Then f is 2-to-1 with $f^{-1}(\{0\}) = \{0\}$. In particular, all Dembowski-Ostrom planar polynomials are 2-to-1.*

Proof. Note that for such a polynomial it holds $f(x) = f(-x)$ for any $x \in \mathbb{F}_{p^n}$. Thus the image set of it contains at most $(p^n + 1)/2$ elements. The rest follows from Theorem 2. □

Corollary 1 covers all known examples of planar mappings listed in the introduction. A natural question arises: Is any planar mapping 2-to-1? The answer is negative as the following observation shows. Given a planar mapping f in $\mathbb{F}_p^n$, $n > 2$, we may choose a linear mapping l such that the planar mapping $f + l$ has more than two zeros. However it might be the case that for any planar mapping f there is a linear mapping l such that $f + l$ is 2-to-1.

In the next proposition we describe a family of planar polynomials which are not 2-to-1. Firstly we recall Wan's upper bound on the image set of polynomials over finite fields. Let g be a polynomial of degree d over a finite field $\mathbb{F}$, which is not a permutation polynomial. Then Wan's bound asserts that $|g(\mathbb{F})| \leq |\mathbb{F}| - (|\mathbb{F}| - 1)/d$, see [12],[11].

Proposition 1. *Let $0 < s < n$ and u be a $(p^s - 2)$th-power in $\mathbb{F}_{p^n}$. Then the planar mapping $f(x) = x^{p^s} - ux^2$ is not 2-to-1 in $\mathbb{F}_{p^n}$.*

Proof. The mapping f is EA-equivalent to x^2 and hence is planar. We will show that there are at least 2 elements $b \in \mathbb{F}_{p^n}$, satisfying $f(x) = f(b)$ if and only if $x = b$. Consider $f(x + b) = f(b)$, which is equivalent to

$$(x + b)^{p^s} - u(x + b)^2 - b^{p^s} + ub^2 = x^{p^s} - 2ubx - ux^2 = 0$$

and thus for $x \neq 0$

$$x^{p^s-1} - ux = 2ub. \tag{4}$$

Note that the mapping $g(x) = x^{p^s-1} - ux = x(x^{p^s-2} - u)$ is not a permutation of $\mathbb{F}_{p^n}$ since $g(x) = 0$ has two solutions. Then using Wan's bound, we get

$$|g(\mathbb{F}_{p^n})| \leq p^n - \frac{p^n - 1}{p^s - 1}.$$

Hence there are at least two elements $b \in \mathbb{F}_{p^n}$ for which (4) has no solution. For such a b it holds $f(b) \neq f(x + b)$ for any $x \neq 0$. □

References

1. Budaghyan, L., Carlet, C., Pott, A.: New classes of almost bent and almost perfect nonlinear polynomials. IEEE Trans. Inform. Theory 52, 1141–1152 (2006)
2. Carlet, C., Charpin, P., Zinoviev, V.: Codes, bent functions and permutations suitable for DES-like cryptosystems. Des. Codes Cryptogr. 15, 125–156 (1998)
3. Coulter, R.S., Henderson, M.: Commutative presemifields and semifields. Adv. Math. 217, 282–304 (2008)

4. Coulter, R.S., Henderson, M., Kosick, P.: Planar polynomials for commutative semifields with specified nuclei. Des. Codes Cryptogr. 44, 275–286 (2007)
5. Coulter, R.S., Matthews, R.W.: Planar functions and planes of Lenz-Barlotti class II. Des. Codes Cryptogr. 10, 167–184 (1997)
6. Dembowski, P., Ostrom, T.: Planes of order n with collineation groups of order n^2. Math. Z. 103, 239–258 (1968)
7. Ding, C., Yin, J.: Signal sets from functions with optimum nonlinearity. IEEE Trans. Communications 55, 936–940 (2007)
8. Ding, C., Yuan, J.: A family of optimal constant-composition codes. IEEE Trans. Inform. Theory 51, 3668–3671 (2005)
9. Ding, C., Yuan, J.: A new family of skew Paley-Hadamard difference sets. J. Comb. Theory Ser. A 113, 1526–1535 (2006)
10. Helleseth, T., Sandberg, D.: Some power mappings with low differential uniformity. Applicable Algebra in Engineering, Communications and Computing 8, 363–370 (1997)
11. Turnwald, G.: A new criterion for permutation polynomials. Finite Fields and Appl. 1, 64–82 (1995)
12. Wan, D.: A p-adic lifting lemma and its applications to permutation polynomials. Finite Fields, Coding Theory and Advances in Comm. and Computing, Lect. Notes in Pure and Appl. Math. 141, 209–216 (1993)
13. Weng, G., Qiu, W., Wang, Z., Xiang, Q.: Pseudo-Paley graphs and skew Hadamard sets from presemifields. Des. Codes Cryptogr. 44, 49–62 (2007)
14. Zha, Z., Kyureghyan, G., Wang, X.: A new family of perfect nonlinear binomials (submitted, 2008)

Classifying 8-Bit to 8-Bit S-Boxes Based on Power Mappings from the Point of DDT and LAT Distributions

Bora Aslan, M. Tolga Sakalli, and Ercan Bulus

Kirklareli University, Computer Tech. and Programming Dept.,
Luleburgaz-Kirklareli, Turkey
Trakya University, Computer Engineering Dept., Edirne, Turkey
Namik Kemal University, Computer Engineering Dept., Corlu-Tekirdag, Turkey
{boraaslan,tolga}@trakya.edu.tr, ercanbulus@corlu.edu.tr

Abstract. S-boxes are vital elements in the design of symmetric ciphers. To date, the techniques for the construction of S-boxes have included pseudo-random generation, finite field inversion, power mappings and heuristic techniques. From these techniques, the use of finite field inversion in the construction of an S-box is so popular because it presents good cryptographic properties. On the other hand, while S-boxes such as AES, Shark, Square and Hierocrypt that are based on inversion mapping over $GF(2^n)$ use an affine transformation after the output of the S-box, in some ciphers like Camellia, an additional affine transformation is used before the input. In this paper, we classify 8-bit to 8-bit S-boxes based on power mappings into classes according to DDT and LAT distributions. Moreover, a formula is given for the calculation of the number of terms in the algebraic expression for a power mapping based S-box according to the given three probable cases.

Keywords: S-boxes, Power Mappings, Classification, DDT, LAT.

1 Introduction

S-boxes are the most important and the only nonlinear component of a block cipher since diffusion and confusion properties which are related with the security of cryptographic algorithms are added to a block cipher by S-boxes. So, bijective S-boxes play an important role in the design of symmetric ciphers. To date, the techniques for the construction of S-boxes have included pseudo-random genera-tion, finite field inversion, power mappings and heuristic techniques. From these techniques, the use of finite field operation in the construction of an S-box yields linear approximation and difference distribution tables in which the entries are close to uniform. Therefore, this provides security against differential and linear attacks. Moreover, because of the fact that S-boxes generated using finite field inversion give good results from the point of cryptographic properties which are LAT (Linear Approximation Table), DDT (Difference Distribution Table - also called XOR Table), completeness, avalanche, strict avalanche, bit independence,

J. von zur Gathen, J.L. Imaña, and Ç.K. Koç (Eds.): WAIFI 2008, LNCS 5130, pp. 123–133, 2008.
© Springer-Verlag Berlin Heidelberg 2008

these algebraic S-boxes like the AES (Advenced Encryption Standard) S-box [1] have received significant attention from cryptographers.

An $n \times n$ S-box, $S(x) : GF(2^n) \rightarrow GF(2^n)$, maps an n-bit input to an n-bit output and can be viewed as consisting of n Boolean functions. This type of S-boxes, one of which is the AES S-box and maps an 8-bit input to an 8-bit output, has been used in most ciphers in the literature.

On the other hand, the Misty 1 [2] and Kasumi [3] which are power mapping based S-boxes can be given as examples of a 7-bit input to a 7-bit output S-boxes and a 9-bit input to a 9-bit output S-boxes respectively. These S-boxes are obtained as linear transforms of power functions over the corresponding fields, with the Kasami's exponent [4]. The Misty 1 S-box and Kasami S-box are obtained $x \rightarrow x^{81}$ in $GF(2^7)$, $x \rightarrow x^5$ in $GF(2^9)$ respectively.

The AES S-box was chosen in terms of Nyberg's suggestion [5] and is based on the inversion mapping over GF (2^n) (with $n = 8$).

$$f(x) = x^{-1}, \ x \in GF(2^8), \ f(0) = 0 \tag{1}$$

As shown in Equation (1), this mapping has a simple algebraic expression that may enable some attacks such as the interpolation attacks [6] [7]. Also, in [6], it is stated that the complexity of such cryptanalytic attacks depends on the degree of the polynomial approximation or the number of terms in the polynomial approximation expression. In order to overcome this problem, this mapping was modified in such a way that does not modify its resistance towards both linear and differential cryptanalysis while overall S-box description becomes complex in $GF(2^8)$. This was achieved by adding a bitwise affine transformation after the inversion mapping [6] [8].

On the other hand, in some ciphers like Camellia [9], an additional affine transformation is used before the input. Therefore, we can define three different probable cases for the place an affine transformation is added. These are:

- case 1 (to add a bitwise affine transformation after the output of an S-box),
- case 2 (to add a bitwise affine transformation before the input of an S-box),
- case 3 (to add a bitwise affine transformation both after the output of an S-box and before the input of an S-box).

A map $f(x)$ from the finite field $GF(p^n)$ to itself is said to be differentially k uniform if k is the maximum number of solutions of the equation $f(x + a) - f(x) = b$ where $a, b \in GF(p^n)$, $a \neq 0$. With $p = 2$, this concept is of interest in cryptography since differential [10] and linear [11] attacks are related to the uniformity of the functions from the point of DDT and LAT. Moreover, while maximum value of b in the equation $f(x+a)+f(x) = b$ (for $p = 2$) where $a \neq 0$, $b \in GF(2^n)$, gives the efficiency of differential cryptanalysis, the maximum LAT value is related to the nonlinearity of the functions. The lower of these two values are, the more resistant function f will be to differential cryptanalysis and linear cryptanalysis. In cryptography, differentially 2 uniform maps are referred to as almost perfect nonlinear maps. Some studies on APN mappings can be found in

Table 1. Known APN functions x^d on $GF(2^n)$, $n = 2m + 1$

Name	Exponent d	ref.
Gold's functions	$2^i + 1$ with $(i, n) = 1$, $1 \leq i \leq m$	[13],[12]
Kasami's functions	$2^{2i} - 2^i + 1$ with $(i, n) = 1$, $2 \leq i \leq m$	[14]
Field inverse	$2^n - 2$	[5]
Welch's function	$2^m + 3$	[15],[16]
Niho's function	$2^m + 2^{m/2} - 1(even\ m)$ $2^m + 2^{(3m+1)/2} - 1(odd\ m)$	[16]
Dobbertin's function	$2^{4i} + 2^{3i} + 2^{2i} + 2^i - 1$ if $n = 5i$	[17]

Table 2. Known APN functions x^d on $GF(2^n)$, $n = 2m$

Name	Exponent d	ref.
Gold's functions	$2^i + 1$ with $(i, n) = 1$, $1 \leq i \leq m$	[13]
Kasami's functions	$2^{2i} - 2^i + 1$ with $(i, n) = 1$, $2 \leq i \leq m$	[14]
Dobbertin's function	$2^{4i} + 2^{3i} + 2^{2i} + 2^i - 1$ if $n = 5i$	[17]

[12][13][14][15][16][17] and according to these studies known APN functions are given in Table 1 and Table 2.

As mentioned before, the AES S-box is based on inversion mapping over $GF(2^8)$ and was chosen in terms of Nyberg suggestion. In fact, inversion mapping over $GF(2^n)$ is a differentially 4 uniform and has the best known nonlinearity [18], that is $2^{n-1} - 2^{n/2}$ [19]. The designers of AES have chosen an S-box which is bijective S-box and fits the byte structure of the cipher. This is because there is no bijective and APN mapping in $GF(2^8)$. So, this encouraged us to focus on other functions with low uniformity from the point of DDT and LAT.

On the other hand, If $x \to x^{127}$ power mapping has been used instead of inversion mapping over $GF(2^8)$, then the algebraic expression of the AES S-box would be

$$S(x) = {}'63' + {}'09'x^{254} + {}'f9'x^{253} + {}'25'x^{251} + {}'f4'x^{247} +$$
$$'01'x^{239} + {}'b5'x^{223} + {}'8f'x^{191} + {}'05'x^{127} \tag{2}$$

while the algebraic expression of the AES S-box is

$$S(x) = {}'63' + {}'05'x^{254} + {}'09'x^{253} + {}'f9'x^{251} + {}'25'x^{247} +$$
$$'f4'x^{239} + {}'01'x^{223} + {}'b5'x^{191} + {}'8f'x^{127}. \tag{3}$$

Note that in Equation (2) and (3) the hexadecimal values in the vertical quote marks represent the field elements in $GF(2^8)$.

The similarity between Equation (2) and (3) encouraged us to classify power functions in $GF(2^8)$ according to the LAT and DDT distributions and have given a clue about classifying power functions according to the degree and number of terms in the algebraic expression. In this paper, we classify power functions in $GF(2^8)$ according to the LAT and DDT distributions. In fact, our study improves findings of Maxwell [20] for $GF(2^8)$.

2   Mathematical Background and Definitions

In this section, we present some basic definitions, propositions, theorems required to classify power functions over $GF(2^8)$.

Definition 1. *Let $S : GF(2^n) \rightarrow GF(2^n)$ be an S-box having an n-bit input and an n-bit output. For any given $a, b \in GF(2^n)$, the XOR table (DDT) can be constructed using*

$$XOR(a, b) = |\{x \in GF(2^n) : S(x) + S(x + a) = b\}| \qquad (4)$$

where a, b are called the input difference and output difference respectively. Also, $\nabla_f = max\{XOR(a, b) : a, b \in GF(2^n), a \neq 0\}$ is called differential uniformity and we say that an S-box is nonlinear if ∇_f smaller than 2^n. Moreover, XOR table of an S-box gives information about the security of the block cipher against differential cryptanalysis. If the differential uniformity is large, this is an indication of an insecure block cipher [21].

Definition 2. *Let $S : GF(2^n) \rightarrow GF(2^n)$ be an S-box having an n-bit input and an n-bit output. For any given $\Gamma_a, \Gamma_b \in GF(2^n)$, the LAT can be constructed using*

$$LAT(\Gamma_a, \Gamma_b) = |\{x \in GF(2^n) : \Gamma_a \bullet x = \Gamma_b \bullet S(x)\}| - 2^{n-1} \qquad (5)$$

where $x \bullet y$ denotes the parity (0 or 1) of bitwise product of x and y. Also, Γ_a, Γ_b are called input mask and output mask respectively. LAT is important tool to measure the security of the S-boxes against linear cryptanalysis. Large elements of LAT are not desired since they indicate high probability of linear relations between the input and output.

Definition 3. *Nonlinearity measure of an $n \times n$ S-box related with the maximum entry of LAT value can be given as*

$$NLM_S = 2^{n-1} - max\,|LAT_S(\Gamma_a, \Gamma_b)| \qquad (6)$$

Definition 4. *Let $f(x) = x^d$ be a function. If $\nabla_f = 2$ for this function, then this function is called APN function.*

Definition 5. *We say that two functions f and g are equivalent if the lists of values $XOR(a, b)$ of these functions with $a, b \in GF(p^n)$ are equal [20].*

Definition 6. *A cyclotomic coset mod N that contains an integer s is the set*

$$C_s = \{s, sq, \ldots, sq^{m-1}\}(mod\ N) \qquad (7)$$

where m is the smallest positive integer such that $sq^m \equiv s\ (mod\ N)$.

Theorem 1. $XOR(a, b)$, where $f(x) = x^d$, is constant on the cyclotomic coset [20].

Proposition 1. Inversion mapping, $f(x) = x^{2^n-2}$ with $x \in GF(2^n)$ over $GF(2^n)$ for n even is differentially 4 uniform [5].

Proposition 2. $f(x) = x^d$ with $x \in GF(2^n)$ over $GF(2^n)$ for n even, where $d = 2^n - 2^i - 1$ for $i = 1, 2, .., n - 1$, is differentially 4 uniform.

Proof. Since $d = x^{2^n-2}$ with $x \in GF(2^n)$ for inversion mapping over $GF(2^n)$, the function $(x)^{(2^n-2).2^i \bmod (2^n-1)}$, according to the Proposition 1 and Theorem 1, is differentially 4 uniform. Therefore,

$$(x)^{(2^n-2).2^i \bmod (2^n-1)} = (x)^{(2^n-1-1).2^i \bmod (2^n-1)}$$

$$= (x)^{(-2^i) \bmod (2^n-1)} \tag{8}$$

means that $f(x) = x^d$ with $x \in GF(2^n)$ where $d = 2^n - 2^i - 1$ for $i = 1, 2, \ldots, n-1$ is differentially 4 uniform.

XOR table and LAT is a table of size $2^8 \times 2^8$ for 8×8 S-boxes whose elements are calculated by (4) and (5) respectively. For the power functions $GF(2^8) \to GF(2^8)$, if $a \neq 0 \in GF(2^8)$ is fixed, and b varies over $GF(2^8)$, then the distribution of $XOR(a, b)$ values is independent of a. Therefore, instead of examining of XOR table size $2^8 \times 2^8$, we can examine table size 1×2^8 and give the distributions of one row of XOR table. Similarly, we can examine one row of LAT table where $\Gamma_a \neq 0 \in GF(2^n)$ is fixed and Γ_b varies over $GF(2^8)$ and give the number of absolute values of LAT elements.

Since all finite fields can be constructed by any irreducible polynomial, we can select any irreducible polynomial to construct the finite field $GF(2^8)$. Hence, as stated in Proposition 1, inversion mapping over $GF(2^8)$ is differentially 4 uniform (the number of 4's, 2's and 0's are 1, 126 and 129 respectively). Moreover, using Proposition 2, we can say that $x \to x^{127}$, $x \to x^{191}$, $x \to x^{223}$, $x \to x^{239}$, $x \to x^{247}$, $x \to x^{251}$, $x \to x^{253}$ power mappings will give the same distributions for DDT as in inversion mapping over $GF(2^8)$.

As mentioned before, since the design of an S-box is generally related with adding affine transformations according to the three probable cases and these transformations are over $GF(2)$, they do not modify cryptographic properties of an S-box but they improve algebraic expression of an S-box. Therefore, we can see power mappings, whether bijective or not, as S-boxes.

3 Classification of Power Functions in $GF(2^8)$

Since all finite fields of the same size are isomorphic, the choice of irreducible polynomial does not make any difference in the construction of the finite field $GF(2^8)$. Therefore, the finite field $GF(2^8)$ has been constructed by using the same irreducible polynomial as in the AES specifications, namely $p(x) = x^8 + +x^4 + x^3 + x + 1$.

Let $\beta = 1 + \alpha$, where α is a root of $p(x)$. Then, we can determine all powers of β (β is a primitive element for this case)

$$\beta^1 = {}'03', \ \beta^2 = {}'05', \ \ldots, \ \beta^{254} = {}'F6', \ \beta^{255} = {}'01'.$$

Table 3. Classification of power functions according to the maximum DDT and the maximum absolute LAT values

Class(d)	Elements of Classes	∇_S	$\lvert N_{Lmax}\rvert$	Nonlinearity Measure of S-boxes ($NLM_s\%$)	
3	(3 6 12 24 48 96 192 129)	2	16	112	(93%)
9	(9 18 36 72 144 33 66 132)	2	16	112	(93%)
39	(39 78 156 57 114 228 201 147)	2	16	112	(93%)
5	(5 10 20 40 80 160 65 130)	4	32	96	(80%)
21	(21 42 84 168 81 162 69 138)	4	16	112	(93%)
95	(95 190 125 150 245 235 215 175)	4	16	112	(93%)
111	(111 222 189 123 246 237 219 183)	4	16	112	(93%)
127	(127 254 253 251 247 239 223 191)	4	16	112	(93%)
7	(7 14 28 56 112 224 193 131)	6	32	96	(80%)
25	(25 50 100 200 145 35 70 140)	6	32	96	(80%)
37	(37 74 148 41 82 164 73 146)	6	32	96	(80%)
63	(63 126 252 249 243 231 207 159)	6	24	104	(87%)
11	(11 22 44 88 176 97 194 133)	10	32	96	(80%)
29	(29 58 116 232 209 163 71 142)	10	32	96	(80%)
13	(13 26 52 104 208 161 67 134)	12	32	96	(80%)
55	(55 110 220 185 115 230 205 155)	12	32	96	(80%)
59	(59 118 236 217 179 103 206 157)	12	32	96	(80%)
15	(15 30 60 120 240 225 195 135)	14	12	116	(97%)
45	(45 90 180 105 210 165 75 150)	14	12	116	(97%)
17	(17 34 68 136)	16	8	120	(100%)
19	(19 38 76 152 49 98 196 137)	16	24	104	(87%)
23	(23 46 92 184 113 226 197 139)	16	32	96	(80%)
31	(31 62 124 248 241 227 199 143)	16	16	112	(93%)
47	(47 94 188 121 242 229 203 151)	16	24	104	(87%)
53	(53 106 212 169 83 166 77 154)	16	32	96	(80%)
61	(61 122 244 233 211 167 79 158)	16	32	96	(80%)
91	(91 182 109 218 181 107 214 173)	16	16	112	(93%)
119	(119 238 221 187)	22	16	112	(93%)
27	(27 54 108 216 177 99 198 141)	26	48	80	(67%)
43	(43 86 172 89 178 101 202 149)	30	48	80	(67%)
87	(87 174 93 186 117 234 213 171)	30	48	80	(67%)
51	(51 102 204 153)	50	12	116	(97%)
85	(85 170)	84	10	118	(98%)
1	(1 2 4 8 16 32 64 128)	256	128	0	(0%)

Table 4. The distribution of the number of probable DDT values for one row

d (class)	The number of x's																		
	0	2	4	6	10	12	14	16	18	22	24	26	28	30	50	52	60	84	256
3	128	128	0	0	0	0	0	0	0	0	0	0	0	0	0	0	0	0	0
9	128	128	0	0	0	0	0	0	0	0	0	0	0	0	0	0	0	0	0
39	128	128	0	0	0	0	0	0	0	0	0	0	0	0	0	0	0	0	0
5	192	0	64	0	0	0	0	0	0	0	0	0	0	0	0	0	0	0	0
21	152	80	24	0	0	0	0	0	0	0	0	0	0	0	0	0	0	0	0
95	156	72	28	0	0	0	0	0	0	0	0	0	0	0	0	0	0	0	0
111	140	104	12	0	0	0	0	0	0	0	0	0	0	0	0	0	0	0	0
127	129	126	1	0	0	0	0	0	0	0	0	0	0	0	0	0	0	0	0
7	157	84	1	14	0	0	0	0	0	0	0	0	0	0	0	0	0	0	0
25	172	48	28	8	0	0	0	0	0	0	0	0	0	0	0	0	0	0	0
37	157	84	1	14	0	0	0	0	0	0	0	0	0	0	0	0	0	0	0
63	156	86	0	14	0	0	0	0	0	0	0	0	0	0	0	0	0	0	0
11	165	66	21	0	4	0	0	0	0	0	0	0	0	0	0	0	0	0	0
29	165	66	21	0	4	0	0	0	0	0	0	0	0	0	0	0	0	0	0
13	149	102	1	0	0	4	0	0	0	0	0	0	0	0	0	0	0	0	0
55	152	96	4	0	0	4	0	0	0	0	0	0	0	0	0	0	0	0	0
59	149	102	1	0	0	4	0	0	0	0	0	0	0	0	0	0	0	0	0
15	134	121	0	0	0	0	1	0	0	0	0	0	0	0	0	0	0	0	0
45	134	121	0	0	0	0	1	0	0	0	0	0	0	0	0	0	0	0	0
17	240	0	0	0	0	0	0	16	0	0	0	0	0	0	0	0	0	0	0
19	159	72	24	0	0	0	0	1	0	0	0	0	0	0	0	0	0	0	0
23	165	60	30	0	0	0	0	1	0	0	0	0	0	0	0	0	0	0	0
31	135	120	0	0	0	0	0	1	0	0	0	0	0	0	0	0	0	0	0
47	159	72	24	0	0	0	0	1	0	0	0	0	0	0	0	0	0	0	0
53	155	96	0	0	0	4	0	1	0	0	0	0	0	0	0	0	0	0	0
61	165	60	30	0	0	0	0	1	0	0	0	0	0	0	0	0	0	0	0
91	135	120	0	0	0	0	0	1	0	0	0	0	0	0	0	0	0	0	0
119	240	0	1	0	2	0	0	9	0	4	0	0	0	0	0	0	0	0	0
27	192	43	0	16	0	4	0	0	0	0	0	1	0	0	0	0	0	0	0
43	185	60	0	8	0	0	0	0	0	0	0	0	1	2	0	0	0	0	0
87	168	85	0	0	0	0	0	0	0	0	0	1	0	2	0	0	0	0	0
51	244	1	0	0	0	0	0	6	0	4	0	0	1	0	0	0	0	0	0
85	252	0	0	0	0	0	0	0	0	0	0	0	0	0	1	2	1	0	0
1	255	0	0	0	0	0	0	0	0	0	0	0	0	0	0	0	0	0	1

Using powers of β, firstly, we have obtained S-boxes according to the power functions, and secondly, obtained DDT and LAT distributions using the definitions in Section 2. In Table 3, the results for ∇_S and maximum absolute value of LAT elements, also denoted as $|N_{Lmax}|$ value, for all classes are shown.

For an $n \times n$ S-box, the maximum nonlinearity measure can be given as $2^n - 2^{n/2-1}$. Therefore, the proportion of nonlinearity measure of an S-box to the maximum nonlinearity measure gives the percentage value of nonlinearity

Table 5. The distribution of the number of probable absolute LAT values for one row

d	The number of $|x|$'s														
(class)	|0|	|2|	|4|	|6|	|8|	|10|	|12|	|14|	|16|	|20|	|24|	|32|	|40|	|48|	|128|
3	65	0	0	0	170	0	0	0	21	0	0	0	0	0	0
9	65	0	0	0	170	0	0	0	21	0	0	0	0	0	0
39	65	0	0	0	170	0	0	0	21	0	0	0	0	0	0
5	49	0	0	0	204	0	0	0	0	0	0	3	0	0	0
21	113	0	0	0	106	0	0	0	37	0	0	0	0	0	0
95	62	0	96	0	36	0	32	0	30	0	0	0	0	0	0
111	49	0	88	0	58	0	40	0	21	0	0	0	0	0	0
127	17	48	36	40	34	24	36	16	5	0	0	0	0	0	0
7	105	0	0	0	120	0	0	0	30	0	0	1	0	0	0
25	115	0	0	0	108	0	0	0	32	0	0	1	0	0	0
37	105	0	0	0	120	0	0	0	30	0	0	1	0	0	0
63	41	0	104	0	72	0	16	0	13	8	2	0	0	0	0
11	101	0	0	0	132	0	0	0	18	0	4	1	0	0	0
29	165	66	21	0	0	4	0	0	0	0	0	0	0	0	0
13	101	0	0	0	132	0	0	0	18	0	4	1	0	0	0
55	99	0	0	0	136	0	0	0	16	0	4	1	0	0	0
59	101	0	0	0	132	0	0	0	18	0	4	1	0	0	0
15	1	0	24	84	85	52	10	0	0	0	0	0	0	0	0
45	1	0	24	84	85	52	10	0	0	0	0	0	0	0	0
17	16	0	0	0	240	0	0	0	0	0	0	0	0	0	0
19	88	0	0	0	152	0	0	0	8	0	8	0	0	0	0
23	90	0	0	0	144	0	0	0	20	0	0	2	0	0	0
31	120	0	0	0	96	0	0	0	40	0	0	0	0	0	0
47	88	0	0	0	152	0	0	0	8	0	8	0	0	0	0
53	60	0	0	0	192	0	0	0	0	0	0	4	0	0	0
61	90	0	0	0	144	0	0	0	20	0	0	2	0	0	0
91	120	0	0	0	96	0	0	0	40	0	0	0	0	0	0
119	16	0	128	0	80	0	0	0	32	0	0	0	0	0	0
27	117	0	0	0	118	0	0	0	16	0	4	0	0	1	0
43	109	0	0	0	136	0	0	0	8	0	0	1	0	2	0
87	109	0	0	0	136	0	0	0	8	0	0	0	2	1	0
51	16	0	64	96	0	64	16	0	0	0	0	0	0	0	0
85	64	0	0	128	0	64	0	0	0	0	0	0	0	0	0
1	255	0	0	0	0	0	0	0	0	0	0	0	0	0	1

measure of an S-box. Also, Table 3 gives nonlinearity measure of S-boxes based on power mappings with the percentage values.

According to Table 3, the classes 3, 9, 39, 5, 21, 95, 111, 25, 63, 55, 15, 45, 27, 85 are not bijective ($gcd(d, 2^8 - 1) \neq 1$) and the classes 3, 9, 39 are APN functions. Although, the classes 5, 21, 95, 111, 127 are differentially 4 uniform, only the class 127 can be used in the design of bijective S-box applications. In addition, the classes 7, 25, 37, 63 are differentially 6 uniform and the classes 7 and 37 give the same DDT distributions where the number of 6's, 4's, 2's, 0's are 14, 1, 84, 157 respectively. For the class 25, the number of 6's, 4's, 2's, 0's are 8, 24, 48, 172 respectively while the number of 6's, 2's, 0's are 14, 86, 156 respectively in the class 63. Detailed description of power functions according to the one row distribution of DDT can found in Table 4.

On the other hand, from the point of LAT distributions, we can say that the classes 3, 9, 39 give the same distribution (the number of 0's, |8|'s, |16|'s are 65,170, 21 respectively). The classes 7 and 37 give the same distribution where the number of 0's, |8|'s, |16|'s and |32|'s are 105, 120, 30, 1 respectively and the number of 0's, |2|'s, |4|'s, |6|'s, |8|'s, |10|'s, |12|'s, |14|'s, |16|'s are 17, 48, 36, 40, 34, 24, 36, 16, 5 respectively in the class 127. One element of this class, that is 254, is used in the AES S-box design.

If we evaluate the nonlinearity measure of S-boxes based on power mappings over $GF(2^8)$ then, we can say that all APN mappings have the nonlinearity measure 112. That means these S-boxes are 93% nonlinear. Moreover, the AES S-box, which is based on the class 127, has also the nonlinearity 112. Generally, there is a parallel relation between the maximum differential value and maximum LAT value for bijective S-boxes. The lower the maximum differential value, the lower the LAT value an S-box has. But, we cannot talk about this relation for non-bijective S-boxes based on power mappings like the class 17 which is not bijective has the maximum nonlinearity with $\nabla_S = 16$. Detailed description of power functions according to the one row distribution of LAT can be found in Table 5.

4   Conclusions

In this paper, we classified 8×8 S-boxes based on power mappings according to the DDT and LAT distributions. For bijective S-boxes based on power mappings, although there are some exceptions, there is a parellel relation between the maximum value of XOR table and maximum absolute value of LAT.

Another important observation is that all elements of a class have the same Hamming weight. Therefore, if any element of a class is used in S-box design, then algebraic weight of this element or this class will affect the number of terms in the algebraic expression according to the cases used in S-box design. For example, if the class 127 is used in the S-box design according to the case 1, 2 and 3, then the number of terms in the algebraic expression will be 9, 255, and 255 respectively with algebraic degree invariable. On the other hand, if the class 7 is used in the S-box design according to the case 1, 2 and 3, then the number of terms in the algebraic expression of these S-boxes will be 9, 93, and 93 respectively. Moreover, the algebraic degree in the algebraic expression of these S-boxes will be 224. A formula for the calculation of the number of terms in the algebraic expression for an S-box designed by using case 2 and 3 can be given as

$$1 + C(n,1) + C(n,2) + \ldots + C(n,r)$$

where r is the Hamming weight of the power function. In addition, the algebraic degree in the algebraic expression of the S-box will be the biggest value among the used class elements.

If case 1 is concerned in the design of an S-box, then all elements of the class used in the S-box design will appear in the algebraic expression and algebraic degree will be the biggest value among the used class elements. So, an improvement of the AES S-box may be considered from the point of the number of terms in the algebraic expression by using case 2 or case 3.

Acknowledgments. The authors would like to thank the anonymous referees for their valuable comments.

References

1. Kavut, S., Yucel, M.D.: On Some Cryptographic Properties of Rijndael. In: Gorodetski, V.I., Skormin, V.A., Popyack, L.J. (eds.) MMM-ACNS 2001. LNCS, vol. 2052, pp. 300–311. Springer, Heidelberg (2001)
2. Matsui, M.: New Block Encryption MISTY. In: Biham, E. (ed.) FSE 1997. LNCS, vol. 1267, pp. 54–68. Springer, Heidelberg (1997)
3. 3rd Generation Partnership Project, Technical Specification Group Services and System Aspects, 3G Security, Specification of the 3GPP Confidentiality and Integrity Algorithms; Document 2: Kasumi Specification, V.3.1.1 (2001)
4. Dobbertin, H.: Almost perfect nonlinear power functions on $GF(2^n)$: the Welch case. IEEE Transactions on Information Theory 45, 1271–1275 (1999)
5. Nyberg, K.: Differentially uniform mappings for cryptography. In: Helleseth, T. (ed.) EUROCRYPT 1993. LNCS, vol. 765, pp. 55–64. Springer, Heidelberg (1994)
6. Jakobsen, T., Knudsen, L.: The interpolation attack on block ciphers. In: Biham, E. (ed.) FSE 1997. LNCS, vol. 1267, pp. 28–40. Springer, Heidelberg (1997)
7. Youssef, A.M., Tavares, S.E., Gong, G.: On Some probabilistic approximations for AES-like s-boxes. Discrete Mathematics 306(16), 2016–2020 (2006)
8. Youssef, A.M., Tavares, S.E.: Affine equivalence in the AES round function. Discrete Applied Mathematics 148(2), 161–170 (2005)
9. Aoki, K., Ichikawa, T., Kanda, M., Matsui, M., Moriai, S., Nakajima, J., Tokita, T.: Camellia: a 128-bit block cipher suitable for multiple platforms-design and analysis. In: Stinson, D.R., Tavares, S. (eds.) SAC 2000. LNCS, vol. 2012, pp. 39–56. Springer, Heidelberg (2001)
10. Biham, E., Shamir, A.: Differential cryptanalysis of DES-like cryptosystems. J.Cryptology 4, 3–72 (1991)
11. Matsui, M.: Linear cryptanalysis method for DES Cipher. In: Helleseth, T. (ed.) EUROCRYPT 1993. LNCS, vol. 765, pp. 386–397. Springer, Heidelberg (1994)
12. Bending, T., Fon-Der- Flaass, D.: Crooked functions, bent functions and distance regular graphs. Electronic Journal of Combinatorics 5:R34, 14 (1998)
13. Gold, R.: Maximal recursive sequences with 3-valued recursive cross-correlation functions. IEEE Transactions on Information Theory 14, 154–156 (1968)
14. Kasami, T.: The weight enumerators for several classes of subcodes of the second order binary Reed-Muller codes. Information and Control 18, 369–394 (1971)
15. Canteaut, A., Charpin, P., Dobbertin, H.: Binary m-sequences with three-valued cross-correlation: a proof of Welch's conjecture. IEEE Transactions on Information Theory 46, 4–8 (2000)
16. Hollman, H.D.L., Xiang, Q.: A proof of the Welch and Niho conjectures on cross-correlations of binary m-sequences. Finite Fields and Their Applications 7, 253–286 (2001)

17. Dobbertin, H.: Almost perfect nonlinear power functions on $GF(2^n)$: a new case for n divisible by 5. In: Jungnickel, D., Niederreiter, H. (eds.) Proceedings of the Conference on Finite Fields and Applications, pp. 113–121. Springer, Berlin (1999)
18. Budaghyan, L., Carlet, C., Felke, P., Leander, G.: An infinite class of quadratic APN functions which are not equvalent to power mappings (2005),
    http://eprint.iacr.org/2005/359.pdf
19. Dobbertin, H.: One to one highly nonlinear power functions on $GF(2^n)$, Applicable Algebra in Engineering. Communication and Computing 9, 139–152 (1998)
20. Maxwell, M.S.: Almost Perfect Nonlinear functions and related combinatorial structures, Phd Thesis, Iowa State University (2005)
21. Akleylek, S., Yucel, M.D.: Comparing Substitution Boxes of the Third Generation GSM and Advanced Encryption Standard Ciphers. In: Information Security and Cryptology Conference, Ankara, Turkey (2007)

EA and CCZ Equivalence of Functions over $GF(2^n)$

K.J. Horadam

RMIT University, Melbourne, VIC 3001, Australia
`kathy.horadam@rmit.edu.au`

Abstract. EA-equivalence classes and the more general CCZ-equivalence classes of functions over $GF(2^n)$ each preserve APN and AB properties desirable for S-box functions. We show that they can be related to subsets $\mathbf{c}[T]$ and $\mathbf{g}[T]$ of equivalence classes $[T]$ of transversals, respectively, thus clarifying their relationship and providing a new approach to their study. We derive a formula which characterises when two CCZ-equivalent functions are EA-inequivalent.

Keywords: CCZ-equivalence, EA-equivalence, bundle, APN function.

1   Introduction

For functions $\phi : G \to N$ between groups, the subset $S_\phi = \{(\phi(x), x) : x \in G\}$ of $N \times G$ can been used as the underlying instrument for measuring the nonlinear behaviour of ϕ under several different measures of nonlinearity that are useful in cryptography and coding. Pott in [13] uses S_ϕ to extend the definition of maximal nonlinearity from the case $N = G = \mathbb{Z}_2^n$ to arbitrary finite abelian groups N and G, in terms of values taken by the group characters of $N \times G$ on S_ϕ. For abelian groups N and G for which $|N|$ divides $|G|$, results of Carlet and Ding [5] show the notions of perfect nonlinearity, bentness and Pott's maximal nonlinearity are equivalent. This generalises the corresponding relationships for functions defined on finite fields.

It remains very difficult to find and classify functions over finite fields that satisfy such desirable nonlinearity conditions, or to determine whether, once found, they are essentially new, that is, inequivalent in some sense to any of the functions already found. Several notions of equivalence exist (c.f. [7, Section 9.2.2]), but the most useful for Boolean functions appear to be *Carlet-Charpin-Zinoviev* (CCZ)-equivalence and *extended affine* (EA)-equivalence.

When $N = G = \mathbb{Z}_2^n$, S_ϕ is called the *graph*[1] of ϕ and is used to define CCZ-equivalence [4,3], which partitions the set of functions into classes with the same nonlinearity and differential uniformity [4, Proposition 3], [3, Proposition 2], but not necessarily the same algebraic degree. EA-equivalent functions have the same

---

[1] In [2,3,4], $\{(x, \phi(x)) : x \in G\}$ is called the graph of ϕ but we swap coordinates for consistency with [13,6,7], without loss of generality.

J. von zur Gathen, J.L. Imaña, and Ç.K. Koç (Eds.): WAIFI 2008, LNCS 5130, pp. 134–143, 2008.
© Springer-Verlag Berlin Heidelberg 2008

nonlinearity, differential uniformity and, for functions of algebraic degree ≥ 2, the same algebraic degree (see [2] for more details).

It is known [4] that EA-equivalence is a particular case of CCZ-equivalence, and that any permutation is CCZ-equivalent to its inverse. In [2], Budaghyan uses the inverse transformation to derive *almost perfect nonlinear* (APN) functions that are EA-inequivalent to any power function, giving the simplest method to construct such functions. Brinkmann and Leander [1] use backtrack programming to classify all the APN functions in dimensions $n = 4$ and $n = 5$. Over $GF(16)$ there is only one CCZ-equivalence class of APN functions, which consists of 2 EA-equivalence classes. Over $GF(32)$ there are 3 CCZ-equivalence classes of APN functions, containing respectively 3, 3 and 1 (for a total of 7) EA-equivalence classes.

However, no general description of how EA-inequivalent functions might partition a CCZ-class is known. This paper is a contribution to this problem.

As a subset of the *group* $E = N \times G$, the graph S_ϕ is a transversal of the normal subgroup $N \times \{1\}$; that is, it intersects each coset $N \times \{x\}$ of $N \times \{1\}$ in E in a single element. Therefore CCZ-equivalence classes should be related in some fashion to equivalence classes of transversals. In earlier work [6,7], the author has related equivalence classes of normalised transversals to equivalence classes of normalised functions $\phi : G \to N$ (called *bundles* $\mathbf{b}(\phi)$) using the theory of group extensions. When $N = G = \mathbb{Z}_2^n$, the bundle equivalence relation between normalised functions $\phi : \mathbb{Z}_2^n \to \mathbb{Z}_2^n$ differs slightly from EA-equivalence, but the author's *affine bundle* $\widehat{\mathbf{b}}(\phi)$ is identical to the EA-equivalence class of ϕ [10, Lemma 1].

Here, restriction to *normalised* functions $\phi : \mathbb{Z}_2^n \to \mathbb{Z}_2^n$ allows us to define a *canonical* transversal T_ϕ and its equivalence class $[T_\phi]$. Inside $[T_\phi]$ is a *canonical* equivalence class $\mathbf{c}[T_\phi]$ of transversals corresponding to the bundle $\mathbf{b}(\phi)$ (Corollary 2). Consideration of the relationship of T_ϕ to its underlying graph S_ϕ leads to the definition of a *graph class* $\mathbf{g}[T_\phi]$ of transversals with $\mathbf{c}[T_\phi] \subseteq \mathbf{g}[T_\phi] \subseteq [T_\phi]$. We introduce the *graph bundle* $\mathcal{B}(\phi)$ of ϕ and show that $\mathbf{b}(\phi) \subseteq \mathcal{B}(\phi)$ (Corollary 2). We relate $\mathcal{B}(\phi)$ to the graph class $\mathbf{g}[T_\phi]$ (Theorem 1), obtaining a version of [3, Proposition 1].

The equivalence relation induced by the $\mathcal{B}(\phi)$ coincides with CCZ-equivalence for normalised functions, and the *affine graph bundle* $\widehat{\mathcal{B}}(\phi)$ containing all translates of functions in $\mathcal{B}(\phi)$ is identical to the CCZ-equivalence class containing ϕ (Lemma 2).

Next (Theorem 3 and Lemma 5) we show that $\varphi \in \mathcal{B}(\phi)$ if and only if there exists $\rho \in \mathrm{Sym}_1(\mathbb{Z}_2^n)$, $s \in \mathbb{Z}_2^n$, a monomorphism $\imath = (\imath_1, \imath_2) : \mathbb{Z}_2^n \to \mathbb{Z}_2^n \times \mathbb{Z}_2^n$ and $\varphi^* \in \mathbf{b}(\phi)$ such that $(\varphi \cdot s) \circ \rho = \imath_1 \circ \varphi^*$ and $\rho = \imath_2 \circ \varphi^*$. In this formula, which characterises how functions in $\mathcal{B}(\phi)$ move away from $\mathbf{b}(\phi)$, the permutation ρ which specifies how a graph underlies a transversal T in $\mathbf{g}[T_\phi]$ seems to be more important than the subgroup $\imath(\mathbb{Z}_2^n)$ of which T is a transversal. In particular, if ρ is an automorphism, $\varphi \in \mathbf{b}(\phi)$ (Theorem 4). If ϕ is itself an automorphism then $\mathrm{inv}(\phi) \in \mathbf{b}(\phi)$ (Example 2).

This gives a new approach to looking for EA-inequivalent APN and other highly nonlinear functions within CCZ equivalence classes as well as for CCZ-inequivalent functions.

The paper is organised as follows. In Section 2 we outline some basic results on graphs and transversals. In Section 3 we show how equivalence of graphs relates to equivalence of transversals and introduce graph classes and graph bundles. In Section 4 we use the theory of group extensions to derive the main results and clarify the relationships between EA and CCZ equivalence classes of normalised functions.

2   Transversals and Graphs

Let G and N be finite abelian groups, written multiplicatively. We denote by $C^1(G, N) = \{f : G \to N, f(1) = 1\}$ the set of all normalised functions from G to N . Any un-normalised f has a *normalisation* $f \cdot 1 \in C^1(G, N)$ given by $f \cdot 1(x) = f(1)^{-1} f(x)$. If f is normalised, then $f \cdot 1 = f$. For $a \in G$, define the *shift action* $f \cdot a$ of a on f by

$$(f \cdot a)(x) = f(a)^{-1} f(ax), \ x \in G. \tag{1}$$

For any $a, b \in G, (f \cdot a) \cdot b = f \cdot (ab)$.

Denote the subgroup of normalised permutations of a group A by $\mathrm{Sym}_1(A)$, the subgroup of automorphisms by $\mathrm{Aut}(A)$ and the inverse of permutation ρ by $\mathrm{inv}(\rho)$. For $\phi \in C^1(G, N)$, define $\partial \phi : G \times G \to N$ to be[2]

$$\partial \phi(x, y) = \phi(x) \phi(y) \phi(xy)^{-1}, \ x, y \in G \tag{2}$$

which measures how much ϕ differs from a homomorphism.

If $N \overset{\iota}{\rightarrowtail} E \overset{\pi}{\twoheadrightarrow} G$ is an extension of N by G (that is, ι and π are group homomorphisms with $\ker \pi = \mathrm{im}\,\iota$) then each section $t : G \to E$ of π (that is, a mapping such that $\pi(t(x)) = x$, $x \in G$) determines a transversal $T = \{t_x = t(x), \ x \in G\}$ of the normal subgroup $\iota(N)$ in E (that is, a set of coset representatives) and vice versa. Every element $e \in E$ has a unique representation as $e = \iota(a) t_x$ for $a \in N$ and $x \in G$. The transversal T is *normalised* if it intersects $\iota(N)$ in 1, or equivalently, if $t_1 = 1$.

For the extension $N \overset{\iota}{\rightarrowtail} N \times G \overset{\kappa}{\twoheadrightarrow} G$, with $\iota(a) = (a, 1)$ and $\kappa(a, x) = x$, and for each $\phi \in C^1(G, N)$, we will call

$$T_\phi = \{t_x = (\phi(x), x), x \in G\} \tag{3}$$

the *canonical transversal of* $\iota(N) = N \times \{1\}$ *in* $N \times G$ *determined by* ϕ. More generally, we study the set underlying T_ϕ.

Definition 1. *Let* $f : G \to N$. *The graph of* f *is the set* $S_f = \{(f(x), x), \ x \in G\} \subset N \times G$. *It is normalised if* $f(1) = 1$.

---

[2] Note that in [6,7] the notation $\partial^{-1}\phi$ is used. The technical reasons for this are irrelevant to our purpose and here we write $\partial \phi$ for simplicity.

This definition is consistent with notation in [13,6,7] and agrees with that in [3, p. 1143] for the case $G = N = (GF(2^n), +) \cong \mathbb{Z}_2^n$, provided we switch first and second components consistently. This can be done with no loss of generality.

The canonical transversal T_ϕ of $N \times \{1\}$ determined by $\phi \in C^1(G, N)$ in (3) has the graph S_ϕ as underlying set, and conversely, the graph S_ϕ of $\phi \in C^1(G, N)$ becomes the canonical transversal T_ϕ in (3) defined by the section $t : x \mapsto t_x = (\phi(x), x), x \in G$. If we know a set is a normalised transversal in $N \times G$, it is easy to identify when it is a translate of a normalised graph, since the translate has a restricted form and hence the translated graph is itself a graph.

Lemma 1. *Suppose $N \overset{\imath}{\rightarrowtail} N \times G \overset{\pi}{\twoheadrightarrow} G$ is an extension and T is a normalised transversal of $\imath(N)$ in $N \times G$. Set $T = \{t_x = (\lambda(x), \rho(x)),\ x \in G\}$, where $\pi(t_x) = x$. The following are equivalent:*

1. *there exists $e \in N \times G$ such that T has the translate eS_ϕ of the graph S_ϕ of $\phi \in C^1(G, N)$ as underlying set;*
2. *there exists $s \in G$ such that T has the translate $(\phi(s), s)^{-1}S_\phi = S_{\phi \cdot s}$ of the graph S_ϕ of $\phi \in C^1(G, N)$ as underlying set;*
3. *$\rho \in \mathrm{Sym}_1(G)$ and there exists $s \in G$ such that $\lambda = (\phi \cdot s) \circ \rho$.* $\square$

If, as in Lemma 1, T is a normalised transversal of $\imath(N)$ with underlying set $S_{\phi \cdot s}$ for $\phi \in C^1(G, N)$ and $s \in G$, we have $\rho \in \mathrm{Sym}_1(G)$ such that $t_x = ((\phi \cdot s) \circ \rho(x), \rho(x)) = (\phi(s), s)^{-1}(\phi \circ (s\rho)(x), (s\rho)(x))$, for $x \in G$.

We represent T by

$$T_{\phi \cdot s}^\rho = \{t_x = ((\phi \cdot s) \circ \rho(x), \rho(x)),\ x \in G\}, \tag{4}$$

where $\pi(t_x) = x$, or, for brevity, by $(\imath, T_\varphi^\rho, \pi)$,   where $\varphi = \phi \cdot s$. Some basic operations on such transversals $(\imath, T_\varphi^\rho, \pi)$ are listed next.

Corollary 1. *Let $\varphi \in C^1(G, N)$, $\rho \in \mathrm{Sym}_1(G)$, and let $(\imath, T_\varphi^\rho, \pi)$ be a transversal as above.*

1. *If $\sigma \in \mathrm{Aut}(G)$, with inverse $\mathrm{inv}(\sigma)$, then $(\imath,\ T_\varphi^{\rho \circ \sigma},\ \mathrm{inv}(\sigma) \circ \pi)$ is a transversal. In particular, if $\rho \in \mathrm{Aut}(G)$ then $(\imath,\ T_\varphi^{\mathrm{id}},\ \rho \circ \pi)$ is a transversal.*
2. *If $\gamma \in \mathrm{Aut}(N)$, then $(\imath \circ \gamma,\ T_\varphi^\rho,\ \pi)$ is a transversal.*

3 Equivalence of Transversals and Graphs

With no loss of generality, we may restrict the study of equivalence of transversals, as defined next, to equivalence of normalised transversals.

Definition 2. *Let T, T' be transversals of the isomorphic normal subgroups K, K', respectively, in a group E. Define T and T' to be equivalent, written $T \sim T'$, if there exist $\alpha \in \mathrm{Aut}(E)$ and $e \in E$ such that $\alpha(K) = K'$ and $eT' = \alpha(T)$, and isomorphic, written $T \cong T'$, if $T' = \alpha(T)$ (ie. $e = 1$). Denote the equivalence class of T by $[T]$.*

From now on, we assume $G = N = \mathbb{Z}_2^n \cong (GF(2^n), +)$, written additively, and $E = \mathbb{Z}_2^n \times \mathbb{Z}_2^n = \mathbb{Z}_2^{2n}$.

We will focus on equivalence classes $[T_\phi]$ for $\phi \in C^1(\mathbb{Z}_2^n, \mathbb{Z}_2^n)$. However, $[T_\phi]$ will usually contain normalised transversals which are not graphs as well a those, such as T_ϕ itself, which are. Consequently, we isolate a subset of $[T_\phi]$ consisting of normalised transversals with (translates of) normalised graphs as underlying set, as well as a special subset containing the canonical transversals.

Definition 3. *Let $\phi \in C^1(\mathbb{Z}_2^n, \mathbb{Z}_2^n)$. Define $\mathbf{c}[T_\phi] \subseteq \mathbf{g}[T_\phi] \subseteq [T_\phi]$ as follows.*

The graph class $\mathbf{g}[T_\phi]$ *of T_ϕ is the set of normalised transversals in $[T_\phi]$ which have a translate of a normalised graph as underlying set; that is, by Lemma 1,*

$$\mathbf{g}[T_\phi] = \{T_{\varphi \cdot s}^\rho : T_{\varphi \cdot s}^\rho \sim T_\phi, \ \varphi \in C^1(\mathbb{Z}_2^n, \mathbb{Z}_2^n), s \in \mathbb{Z}_2^n, \rho \in \mathrm{Sym}_1(\mathbb{Z}_2^n)\}.$$

The canonical class $\mathbf{c}[T_\phi]$ *of T_ϕ is the set of canonical transversals in $[T_\phi]$; that is,*

$$\mathbf{c}[T_\phi] = \{T_\varphi : T_\varphi \sim T_\phi, \ \varphi \in C^1(\mathbb{Z}_2^n, \mathbb{Z}_2^n)\}.$$

By Corollary 1 it is possible that the graph class $\mathbf{g}[T_\phi]$ may contain transversals $(\imath, T_\phi^{\mathrm{id}}, \pi)$ with exactly the same elements as the canonical transversal $(\imath, T_\phi, \kappa)$, that is, defined by the same section, which are transversal to some normal subgroup $\imath(N)$ different from $\iota(N) = \mathbb{Z}_2^n \times \{0\}$. An example to show that this does occur is given in [8].

We now introduce an equivalence relation on normalised functions which coincides with CCZ-equivalence by affine permutations. This permits us to relate CCZ and EA equivalence very naturally using transversals. Because this equivalence relation is defined for functions between arbitrary groups in [8], we use a different name.

Definition 4. *Two functions $\phi, \varphi \in C^1(\mathbb{Z}_2^n, \mathbb{Z}_2^n)$ are* graph equivalent *if there exist $\alpha \in \mathrm{Aut}(\mathbb{Z}_2^n \times \mathbb{Z}_2^n)$ and $e \in \mathbb{Z}_2^n \times \mathbb{Z}_2^n$ such that $\alpha(S_\phi) = e + S_\varphi$. They are* graph isomorphic *if $\alpha(S_\phi) = S_\varphi$, ie. $e = 0$. Denote the graph equivalence class of ϕ by $\mathcal{B}(\phi)$ and term it the* graph bundle *of ϕ.*

Two functions $f, f' : \mathbb{Z}_2^n \to \mathbb{Z}_2^n$ are affine graph equivalent *if their normalisations $f \cdot 0, f' \cdot 0 \in C^1(\mathbb{Z}_2^n, \mathbb{Z}_2^n)$ are graph equivalent; that is, if there exist $\alpha \in \mathrm{Aut}(\mathbb{Z}_2^n \times \mathbb{Z}_2^n)$ and $e \in \mathbb{Z}_2^n \times \mathbb{Z}_2^n$ such that $\alpha(S_{f \cdot 0}) = e + S_{f' \cdot 0}$. Denote the* affine graph equivalence class *of f by $\widehat{\mathcal{B}}(f)$ and term it the* affine graph bundle *of f.*

Because $S_f = (f(0), 0) + S_{f \cdot 0}$ we see that

$$f \in \widehat{\mathcal{B}}(f') \Leftrightarrow f \cdot 0 \in \mathcal{B}(f' \cdot 0), \tag{5}$$

so again, we may restrict to normalised functions with no loss of generality.

We now show that affine graph equivalence equals CCZ-equivalence (using the *affine* permutation definition of CCZ-equivalence in [3, Definition 1]). The *linear* permutation case of CCZ-equivalence corresponds to a particular case of affine graph equivalence, which for normalised functions is graph *isomorphism*.

Lemma 2. *Let $f, f' : \mathbb{Z}_2^n \to \mathbb{Z}_2^n$. Then f is CCZ-equivalent to f' if and only if $\widehat{\mathcal{B}}(f) = \widehat{\mathcal{B}}(f')$.*

Proof. By (5), $f \in \widehat{\mathcal{B}}(f')$ if and only if there exist $\alpha \in \mathrm{Aut}(\mathbb{Z}_2^n \times \mathbb{Z}_2^n)$ and $e \in \mathbb{Z}_2^n \times \mathbb{Z}_2^n$ such that $\alpha(S_f) = \alpha((f(0), 0)) + e - (f'(0), 0) + S_{f'}$ if and only if there exist $\alpha \in \mathrm{Aut}(\mathbb{Z}_2^n \times \mathbb{Z}_2^n)$ and $e' \in \mathbb{Z}_2^n \times \mathbb{Z}_2^n$ such that $\alpha(S_f) = e' + S_{f'}$ if and only if f is CCZ-equivalent to f'. Note $e' = 0$ if and only if $e = (f'(0), 0) - \alpha((f(0), 0))$. $\square$

Since a graph is always representable as a canonical transversal, we may describe graph equivalence in terms of graphs and transversals. The affine version of Theorem 1.1 is the extension of the characterisation [3, Proposition 1] of CCZ-equivalence from the linear permutation case to the affine permutation case.

Theorem 1. *Let $\phi, \varphi \in C^1(\mathbb{Z}_2^n, \mathbb{Z}_2^n)$. Let S_ϕ, S_φ, be their respective graphs and let T_ϕ be the canonical transversal (3) determined by ϕ.*

1. *$\varphi \in \mathcal{B}(\phi)$ if and only if there exist $\alpha \in \mathrm{Aut}(\mathbb{Z}_2^n \times \mathbb{Z}_2^n)$, $\rho \in \mathrm{Sym}_1(\mathbb{Z}_2^n)$ and $s \in \mathbb{Z}_2^n$ such that*

$$\alpha(T_\phi) = T_{\varphi \cdot s}^\rho ;$$

2. *$\varphi \in \mathcal{B}(\phi)$ if and only if there exist $s \in \mathbb{Z}_2^n$ and $\rho \in \mathrm{Sym}_1(\mathbb{Z}_2^n)$ such that $T_{\varphi \cdot s}^\rho \in \mathbf{g}[T_\phi]$.*

Proof. 1. By definition, $\varphi \in \mathcal{B}(\phi)$ if and only if there exist $\alpha \in \mathrm{Aut}(\mathbb{Z}_2^n \times \mathbb{Z}_2^n)$ and $e \in \mathbb{Z}_2^n \times \mathbb{Z}_2^n$ such that the normalised transversal $\alpha(T_\phi)$ of $\alpha(\mathbb{Z}_2^n \times \{0\})$ (with underlying set $\alpha(S_\phi)$) has the translate $e + S_\varphi$ as underlying set. Then Lemma 1 applies.

2. By Part 1 and Lemma 1, if $\varphi \in \mathcal{B}(\phi)$ there exist $s \in \mathbb{Z}_2^n$ and $\rho \in \mathrm{Sym}_1(\mathbb{Z}_2^n)$ such that $T_{\varphi \cdot s}^\rho \cong T_\phi$, so $T_{\varphi \cdot s}^\rho \in \mathbf{g}[T_\phi]$. Conversely if there exist $s \in \mathbb{Z}_2^n$ and $\rho \in \mathrm{Sym}_1(\mathbb{Z}_2^n)$ such that $T_{\varphi \cdot s}^\rho \sim T_\phi$, then there exist $\alpha \in \mathrm{Aut}(\mathbb{Z}_2^n \times \mathbb{Z}_2^n)$ and $e \in \mathbb{Z}_2^n \times \mathbb{Z}_2^n$ such that $\alpha(T_\phi) = e + T_{\varphi \cdot s}^\rho$, so $\alpha(S_\phi) = e - (\varphi(s), s) + S_\varphi$ and $\varphi \in \mathcal{B}(\phi)$. $\square$

In the next section, we use the theory of group extensions to relate graph bundles $\mathcal{B}(\phi)$ to bundles $\mathbf{b}(\phi)$.

4   Transversals and Bundles

Transversals T are used in the theory of group extensions to define cocycles ψ_T according to the following standard construction. See [7] for further details.

Lemma 3. *Suppose that $\mathbb{Z}_2^n \overset{\imath}{\rightarrowtail} \mathbb{Z}_2^n \times \mathbb{Z}_2^n \overset{\pi}{\twoheadrightarrow} \mathbb{Z}_2^n$ is an extension of $\mathbb{Z}_2^n$ by $\mathbb{Z}_2^n$ and let $T = \{t_x, x \in \mathbb{Z}_2^n : \pi(t_x) = x\}$ be a normalised transversal of $\imath(\mathbb{Z}_2^n)$ in $\mathbb{Z}_2^n \times \mathbb{Z}_2^n$. Then ψ_T defined by*

$$\psi_T(x, y) = \imath^{-1}(t_x + t_y - t_{xy}), \tag{6}$$

for all $x, y \in \mathbb{Z}_2^n$, is a cocycle, and must be of the form $\partial\phi$ for some $\phi \in C^1(\mathbb{Z}_2^n, \mathbb{Z}_2^n)$. $\square$

The mapping from transversal to cocycle given in Lemma 3 is surjective. We illustrate this for the case at hand.

Lemma 4. *Let* $\phi \in C^1(\mathbb{Z}_2^n, \mathbb{Z}_2^n)$. *Let* T_ϕ *in (3) be the canonical transversal. Then* $\psi_{T_\phi} = \partial \phi$. $\qquad\qquad\qquad\square$

Equivalence of transversals determines a corresponding equivalence of cocycles.

Theorem 2. [7, Theorem 8.5] *Let* T *and* T' *be normalised transversals in* $\mathbb{Z}_2^n \times \mathbb{Z}_2^n = E$ *of the normal subgroups* K *and* K' *isomorphic to* $\mathbb{Z}_2^n$, *respectively, for which* $E/K \cong E/K' \cong \mathbb{Z}_2^n$. *Let* $\psi_T = \partial\phi$, $\psi_{T'} = \partial\varphi$ *be the corresponding cocycles of Lemma 3, respectively. The following are equivalent:*

1. $T \sim T'$;
2. *there exist* $\gamma, \theta \in Aut(\mathbb{Z}_2^n)$ *and* $a \in \mathbb{Z}_2^n$ *such that*

$$\partial\varphi = \partial(\gamma \circ (\phi \cdot a) \circ \theta));\qquad\qquad(7)$$

3. *there exist* $\gamma, \theta \in Aut(\mathbb{Z}_2^n)$, $a \in \mathbb{Z}_2^n$ *and* $\chi \in \mathrm{Hom}(\mathbb{Z}_2^n, \mathbb{Z}_2^n)$ *such that*

$$\varphi = (\gamma \circ (\phi \cdot a) \circ \theta)\, \chi.\qquad\qquad\square$$

Equivalence classes of normalised functions $\phi : \mathbb{Z}_2^n \to \mathbb{Z}_2^n$ are defined using Theorem 2.3. These equivalence classes are termed bundles; that is, the **bundle** $\mathbf{b}(\phi)$ of ϕ is

$$\left\{ (\gamma \circ (\phi \cdot a) \circ \theta)\, \chi : (\gamma, a, \theta, \chi) \in (\mathrm{Aut}(\mathbb{Z}_2^n), \mathbb{Z}_2^n, \mathrm{Aut}(\mathbb{Z}_2^n), \mathrm{Hom}(\mathbb{Z}_2^n, \mathbb{Z}_2^n)) \right\}\quad(8)$$

Bundles exactly characterise the canonical classes of transversals.

Corollary 2. *Let* $\phi, \varphi \in C^1(\mathbb{Z}_2^n, \mathbb{Z}_2^n)$.

1. [7, Theorem 9.22] $\mathbf{b}(\phi) = \mathbf{b}(\varphi) \Leftrightarrow \mathbf{c}[T_\phi] = \mathbf{c}[T_\varphi]$. *Hence* $\mathbf{c}[T_\phi] =$

$$\left\{ T_{(\gamma\circ(\phi\cdot a)\circ\theta)\chi} : (\gamma, a, \theta, \chi) \in (\mathrm{Aut}(\mathbb{Z}_2^n), \mathbb{Z}_2^n, \mathrm{Aut}(\mathbb{Z}_2^n), \mathrm{Hom}(\mathbb{Z}_2^n, \mathbb{Z}_2^n)) \right\}.\quad(9)$$

2. $\mathbf{b}(\phi) \subseteq \mathcal{B}(\phi)$ *so if* $\mathbf{b}(\varphi) = \mathbf{b}(\phi)$ *then* $\mathcal{B}(\varphi) = \mathcal{B}(\phi)$ *and* $\mathbf{c}[T_\phi] = \mathbf{c}[T_\varphi] \subset \mathbf{g}[T_\varphi] \cap \mathbf{g}[T_\phi]$.

Proof. 1. By definition, $\mathbf{c}[T_\phi] = \mathbf{c}[T_\varphi] \Leftrightarrow T_\phi \sim T_\varphi$ which, by Lemma 4, Theorem 2 and (8), holds $\Leftrightarrow \mathbf{b}(\phi) = \mathbf{b}(\varphi)$.

2. By Part 1, if $\varphi \in \mathbf{b}(\phi)$, $T_\phi \sim T_\varphi$ and by Definition 2 there exist $\alpha \in \mathrm{Aut}(\mathbb{Z}_2^n \times \mathbb{Z}_2^n)$ and $e \in \mathbb{Z}_2^n \times \mathbb{Z}_2^n$ such that $\alpha(\mathbb{Z}_2^n \times \{0\}) = \mathbb{Z}_2^n \times \{0\}$ and $\alpha(T_\phi) = e + T_\varphi$, so $\alpha(S_\phi) = e + S_\varphi$. By Definition 4, $\varphi \in \mathcal{B}(\phi)$. The rest follows by symmetry. $\qquad\qquad\square$

The *affine bundle* $\widehat{\mathbf{b}}(f)$ of $f : \mathbb{Z}_2^n \to \mathbb{Z}_2^n$ is

$$\widehat{\mathbf{b}}(f) = \{ f' : \mathbb{Z}_2^n \to \mathbb{Z}_2^n,\ f' \cdot 0 \in \mathbf{b}(f \cdot 0) \}.\qquad\qquad(10)$$

Example 1. [10, Lemma 1] The affine bundle $\widehat{\mathbf{b}}(f)$ of $f : \mathbb{Z}_2^n \to \mathbb{Z}_2^n$ equals the EA-equivalence class of f.

In general, the surjective mapping of Lemma 4 is not injective: more than one transversal will define the same cocycle. All transversals determining the same cocycle $\partial\phi$ as T_ϕ may be characterised, using Lemma 3.

Lemma 5. *Let* $\mathbb{Z}_2^n \overset{\imath}{\rightarrowtail} \mathbb{Z}_2^n \times \mathbb{Z}_2^n \overset{\pi}{\twoheadrightarrow} \mathbb{Z}_2^n$ *be an extension of* $\mathbb{Z}_2^n$ *by* $\mathbb{Z}_2^n$. *Let* $T = \{t_x,\ x \in \mathbb{Z}_2^n\}$ *with* $\pi(t_x) = x$ *be a normalised transversal in* $\mathbb{Z}_2^n \times \mathbb{Z}_2^n$ *of the normal subgroup* $K = \imath(\mathbb{Z}_2^n)$, *with corresponding cocycle* ψ_T. *Set* $\imath(n) = (\imath_1(n), \imath_2(n))$, $n \in \mathbb{Z}_2^n$, *for* $\imath_1, \imath_2 \in \mathrm{Hom}(\mathbb{Z}_2^n, \mathbb{Z}_2^n)$, *and* $t_x = (\lambda(x), \rho(x))$, $x \in \mathbb{Z}_2^n$ *for* $\lambda,\ \rho \in C^1(\mathbb{Z}_2^n, \mathbb{Z}_2^n)$. *Let* $\phi \in C^1(\mathbb{Z}_2^n, \mathbb{Z}_2^n)$.
Then $\psi_T = \partial\phi$ *if and only if there exists* $\chi \in \mathrm{Hom}(\mathbb{Z}_2^n, \mathbb{Z}_2^n)$ *such that*

$$\lambda = \imath_1 \circ (\phi + \chi), \qquad \rho = \imath_2 \circ (\phi + \chi). \tag{11}$$

In particular, if $\varphi \in C^1(\mathbb{Z}_2^n, \mathbb{Z}_2^n)$, $s \in \mathbb{Z}_2^n$ *and* $\rho \in \mathrm{Sym}_1(\mathbb{Z}_2^n)$, $\psi_{T_{\varphi \cdot s}} = \partial\phi$ *if and only if there exist* $\chi \in \mathrm{Hom}(\mathbb{Z}_2^n, \mathbb{Z}_2^n)$ *such that*

$$(\varphi \cdot s) \circ \rho = \imath_1 \circ (\phi + \chi), \qquad \rho = \imath_2 \circ (\phi + \chi). \tag{12}$$

Proof. Note $\imath_1(\mathbb{Z}_2^n)$, $\imath_2(\mathbb{Z}_2^n)$ are subgroups of $\mathbb{Z}_2^n$. Application of Lemma 3 is straightforward:

$$\begin{aligned}
\imath(\psi_T(x, y)) &= t_x + t_y - t_{xy} \\
&= (\lambda(x) + \lambda(y) - \lambda(xy),\ \rho(x) + \rho(y) - \rho(xy)) \\
&= (\partial\lambda(x, y), \partial\rho(x, y))
\end{aligned}$$

by (2), so $\psi_T = \partial\phi \Leftrightarrow \imath \circ \psi_T = \imath \circ \partial\phi \Leftrightarrow (\partial\lambda, \partial\rho) = (\imath_1 \circ \partial\phi, \imath_2 \circ \partial\phi) \Leftrightarrow (\partial\lambda, \partial\rho) = (\partial(\imath_1 \circ \phi), \partial(\imath_2 \circ \phi))$, because $\imath_1$ and $\imath_2$ are homomorphisms. $\qquad\square$

Next we relate bundles and graph bundles. If S_φ is a transversal T in $[T_\phi]$ this DOES NOT necessarily imply $\varphi \in \mathbf{b}(\phi)$ (though this can be true). Nor does it imply that $\psi_T = \partial\varphi$, though again, this may be true. To combine Theorem 2 with Theorem 1 we need the case $\rho \in \mathrm{Sym}_1(\mathbb{Z}_2^n)$ of Lemma 5.

Theorem 3. *Let* $\phi, \varphi \in C^1(\mathbb{Z}_2^n, \mathbb{Z}_2^n)$. *Then* $\varphi \in \mathcal{B}(\phi)$ *if and only if there exist* $\rho \in \mathrm{Sym}_1(\mathbb{Z}_2^n)$, $s \in \mathbb{Z}_2^n$, θ, $\gamma \in \mathrm{Aut}(\mathbb{Z}_2^n)$ *and* $\chi \in \mathrm{Hom}(\mathbb{Z}_2^n, \mathbb{Z}_2^n)$ *such that*

$$\psi_{T_{\varphi \cdot s}^\rho} = \partial\varphi^* \tag{13}$$

for

$$\varphi^* = (\gamma \circ \phi \circ \theta) + \chi.$$

Proof. By Theorem 2, $\varphi \in \mathcal{B}(\phi)$ if and only if there exist $\rho \in \mathrm{Sym}_1(\mathbb{Z}_2^n)$ and $s \in \mathbb{Z}_2^n$ such that $T_{\varphi \cdot s}^\rho \cong T_\phi$, if and only if (by Lemma 4 and Theorem 2 with $e = 0$) there exist $\rho \in \mathrm{Sym}_1(\mathbb{Z}_2^n)$, $s \in \mathbb{Z}_2^n$, $\theta \in \mathrm{Aut}(\mathbb{Z}_2^n)$, $\gamma \in \mathrm{Aut}(\mathbb{Z}_2^n)$ and $\chi \in \mathrm{Hom}(\mathbb{Z}_2^n, \mathbb{Z}_2^n)$ such that $\psi_{T_{\varphi \cdot s}^\rho} = \partial\varphi^*$ where $\varphi^* = (\gamma \circ \phi \circ \theta) + \chi$. $\qquad\square$

From Theorem 3 and Lemma 5 we see that $\varphi \in \mathcal{B}(\phi)$ if and only if there exists $\rho \in \mathrm{Sym}_1(\mathbb{Z}_2^n)$, $s \in \mathbb{Z}_2^n$, a monomorphism $\imath = (\imath_1, \imath_2) : \mathbb{Z}_2^n \to \mathbb{Z}_2^n \times \mathbb{Z}_2^n$ and $\varphi^* \in \mathbf{b}(\phi)$ such that $(\varphi \cdot s) \circ \rho = \imath_1 \circ \varphi^*$ and $\rho = \imath_2 \circ \varphi^*$.

An important example is the case $\rho \in \mathrm{Aut}(\mathbb{Z}_2^n)$.

Theorem 4. *Let* $\phi, \varphi \in C^1(\mathbb{Z}_2^n, \mathbb{Z}_2^n)$ *and* $\varphi \in \mathcal{B}(\phi)$. *If there exist* $\rho \in \mathrm{Aut}(\mathbb{Z}_2^n)$ *and* $s \in \mathbb{Z}_2^n$ *such that* $T_{\varphi \cdot s}^\rho \cong T_\phi$ *then* $\varphi \in \mathbf{b}(\phi)$.

Proof. If $T_{\varphi \cdot s}^\rho \cong T_\phi$, there exists $\alpha \in \mathrm{Aut}(\mathbb{Z}_2^n \times \mathbb{Z}_2^n)$ such that $T_{\varphi \cdot s}^\rho = \alpha(T_\phi)$ is a normalised transversal of $\alpha \circ \iota(\mathbb{Z}_2^n)$, where $\alpha(t_x) = ((\varphi \cdot s) \circ \rho(x), \rho(x))$ for $t_x \in T_\phi$. By Lemma 3, $\alpha \circ \iota(\psi_{T_{\varphi \cdot s}^\rho}(x, y)) = \alpha \circ \iota(\partial \phi(x, y))$; that is, $\psi_{T_{\varphi \cdot s}^\rho} = \partial \phi$. Let $J = (\mathbb{Z}_2^n \times \{0\}) \cap \alpha(\mathbb{Z}_2^n \times \{0\})$, so α restricted to J is an automorphism which extends to an automorphism of $\mathbb{Z}_2^n \times \{0\}$. Thus there is a $\gamma \in \mathrm{Aut}(\mathbb{Z}_2^n)$ such that $\alpha \circ \iota = \iota \circ \gamma$ on J. Because ρ is a homomorphism, $\alpha \circ \iota(\psi_{T_{\varphi \cdot s}^\rho}(x, y)) = \big(\partial((\varphi \cdot s) \circ \rho)(x, y), 0\big) = \iota(\partial((\varphi \cdot s) \circ \rho)(x, y)) = \iota \circ \gamma(\partial \phi(x, y))$. Therefore $\partial((\varphi \cdot s) \circ \rho) = \partial(\gamma \circ \phi)$. By Theorem 2, $\varphi \in \mathbf{b}(\phi)$. $\qquad \square$

In [7, Corollary 9.23], the author claimed incorrectly that a normalised permutation and its inverse lie in the same bundle. For instance, a Gold power function $\phi(x) = x^{2^i+1}, (i, n) = 1$, over $G = (GF(2^n), +)$, n odd, has algebraic degree 2, so is not itself affine, and thus all functions in its bundle have algebraic degree 2 [3, p. 1142]. However its inverse has algebraic degree $(n + 1)/2$ [11]. Instead, a normalised permutation and its inverse lie in the same graph bundle.

Example 2. Let $\phi \in \mathrm{Sym}_1(\mathbb{Z}_2^n)$ have inverse $\mathrm{inv}(\phi)$. Then

1. $\mathcal{B}(\mathrm{inv}(\phi)) = \mathcal{B}(\phi)$;
2. if $\phi \in \mathrm{Aut}(\mathbb{Z}_2^n)$, $\mathbf{b}(\phi) = \mathbf{b}(\mathrm{inv}(\phi))$.

Proof. 1. Consider the extension $\mathbb{Z}_2^n \xrightarrow{\imath} \mathbb{Z}_2^n \times \mathbb{Z}_2^n \xrightarrow{\pi} \mathbb{Z}_2^n$, with $\imath(y) = (0, y)$ and $\pi((x, y)) = x$. The transversal $T = \{t_x^\phi = (x, \phi(x)), x \in \mathbb{Z}_2^n\}$ with $\pi(t_x^\phi) = x$ of $\{0\} \times \mathbb{Z}_2^n$ in $\mathbb{Z}_2^n \times \mathbb{Z}_2^n$ is $T_{\mathrm{inv}(\phi)}^\phi$ by (4), and $\psi_T = \partial \phi$, by Lemma 3. Since $\phi \in \mathbf{b}(\phi)$ the result follows from Theorem 3 on setting $\varphi^* = \rho = \phi$ and $\varphi = \mathrm{inv}(\phi)$.

2. If $\phi \in \mathrm{Aut}(\mathbb{Z}_2^n)$, $\mathrm{inv}(\phi) = \mathrm{inv}(\phi) \circ \phi \circ \mathrm{inv}(\phi)$ satisfies (8). $\qquad \square$

In order to identify the way in which $\mathcal{B}(\phi)$ partitions into bundles (see Corollary 2) we must first isolate the $\rho \in \mathrm{Sym}_1(\mathbb{Z}_2^n)$ and $\varphi \in \mathcal{B}(\phi)$ such that $\varphi \in \mathbf{b}(\phi)$. By Theorem 4 the set of such ρ includes $\mathrm{Aut}(\mathbb{Z}_2^n)$. We conclude that it is the set of permutations ρ for which there is a transversal $T_{\varphi \cdot s}^\rho$ which is the key to these problems, rather than the way in which automorphisms of $\mathbb{Z}_2^n \times \mathbb{Z}_2^n$ act on the subgroup $\mathbb{Z}_2^n \times \{0\}$.

Acknowledgements. The author is very grateful to Claude Carlet, Lilya Budaghyan and Alex Pott for numerous conversations and insightful comments which helped clarify her understanding of EA and CCZ equivalence, and correct errors in earlier publications [6,7,9]. She is also grateful to Pascale Charpin and the hospitality of INRIA, where this work was initiated. Finally she is grateful to the three anonymous referees whose comments much improved the clarity and motivation of the exposition.

References

1. Brinkmann, M., Leander, G.: On the classification of APN functions up to dimension 5. In: Augot, D., Sendrier, N., Tillich, J.-P. (eds.) Proc. International Workshop on Coding and Cryptography, Versailles, France, pp. 39–58 (2007)
2. Budaghyan, L.: The simplest method for constructing APN polynomials EA-inequivalent to power functions. In: Carlet, C., Sunar, B. (eds.) WAIFI 2007. LNCS, vol. 4547, pp. 177–188. Springer, Heidelberg (2007)
3. Budaghyan, L., Carlet, C., Pott, A.: New classes of almost bent and almost perfect nonlinear polynomials. IEEE Trans. Inform. Theory 52, 1141–1152 (2006)
4. Carlet, C., Charpin, P., Zinoviev, V.: Codes, bent functions and permutations suitable for DES-like cryptosystems. Des. Codes Cryptogr. 15, 125–156 (1998)
5. Carlet, C., Ding, C.: Highly nonlinear mappings. J. Complexity 20, 205–244 (2004)
6. Horadam, K.J.: A theory of highly nonlinear functions. In: Fossorier, M., Imai, H., Lin, S., Poli, A. (eds.) AAECC 2006. LNCS, vol. 3857, pp. 87–100. Springer, Heidelberg (2006)
7. Horadam, K.J.: Hadamard Matrices and Their Applications. Princeton University Press, Princeton (2007)
8. Horadam, K.J.: Transversals, graphs and bundles of functions, in preparation.
9. Horadam, K.J., Farmer, D.G.: Bundles, presemifields and nonlinear functions, extended abstract. In: Augot, D., Sendrier, N., Tillich, J.-P. (eds.) Proc. International Workshop on Coding and Cryptography, Versailles, France, pp. 197–206 (2007)
10. Horadam, K.J., Farmer, D.G.: Bundles, presemifields and nonlinear functions. Des., Codes Cryptogr. (to appear, 2008), doi:10.1007/s10623-008-9172-z
11. Nyberg, K.: Differentially uniform mappings for cryptography. In: Helleseth, T. (ed.) EUROCRYPT 1993. LNCS, vol. 765, pp. 55–64. Springer, Heidelberg (1994)
12. Pott, A.: A survey on relative difference sets. In: Groups, Difference Sets and the Monster, pp. 195–232. de Gruyter, New York (1996)
13. Pott, A.: Nonlinear functions in abelian groups and relative difference sets. Discr. Appl. Math. 138, 177–193 (2004)

On the Number of Two-Weight Cyclic Codes with Composite Parity-Check Polynomials

Gerardo Vega

Dirección General de Servicios de Cómputo Académico, Universidad Nacional
Autónoma de México, 04510 México D.F., Mexico
gerardov@servidor.unam.mx

Abstract. Sufficient conditions for the construction of a two-weight cyclic code by means of the direct sum of two one-weight cyclic codes, were recently presented in [4]. On the other hand, an explicit formula for the number of one-weight cyclic codes, when the length and dimension are given, was proved in [3]. By imposing some conditions on the finite field, we now combine both results in order to give a lower bound for the number of two-weight cyclic codes with composite parity-check polynomials.

Keywords: One-weight cyclic codes and two-weight cyclic codes.

1 Introduction

In [5] it was proved that if C is a two-weight projective cyclic code of dimension k over $\mathbb{F}_q$, then either

1) C is irreducible, or
2) if $q \neq 2$, C is the direct sum of two one-weight cyclic codes of length $n = \lambda(\frac{q^k-1}{q-1})$ where λ divides $q - 1$ and $\lambda \neq 1$. Additionally, the two nonzero weights of C are $(\lambda - 1)q^{k-1}$ and λq^{k-1} (direct sum here means direct sum as vector spaces).

An infinite class of two-weight cyclic projective codes which are irreducible is known for any q (deduced from semi-primitive codes). The propose of this work, is to show that it is also an infinite class for the second case. That is, we want to determine the number of two-weight (projective or not projective) cyclic codes that are constructed by means of the direct sum of two one-weight cyclic codes of the same length and dimension. For one-weight cyclic codes it is known (see [3]) that, for a given length and dimension, it is possible to determine the exact number of this kind of codes. Using this and the sufficient conditions for the construction of a two-weight cyclic code by means of the direct sum of two one-weight cyclic codes, that were recently presented in [4], we now give, for some finite fields, a lower bound for the number of two-weight cyclic codes with composite parity-check polynomials having exactly two irreducible factors of the same degree.

J. von zur Gathen, J.L. Imaña, and Ç.K. Koç (Eds.): WAIFI 2008, LNCS 5130, pp. 144–156, 2008.

Since the main components of this work are the one-weight cyclic codes, it is important to keep in mind the restriction on the length of this kind of codes. For this, let C be an $[n, k]$ linear code over $\mathbb{F}_q$, whose dual weight is at least 2. It is well known (see for example [5]) that if, additionally, C is a one-weight code, then its length n must be given by

$$n = \lambda \frac{q^k - 1}{q - 1} , \tag{1}$$

for some positive integer λ. Since the minimal distance of the dual of any nonzero cyclic code over $\mathbb{F}_q$ is greater than 1, it follows that the length n of all one-weight cyclic codes of dimension k, is given by (1). Here we are going to assume $n \leq q^k - 1$. Thus, since all the one-weight cyclic codes are irreducible, then, for this particular case, the zeros of $x^n - 1$, which form a cyclic group, lie in the extension field $\mathbb{F}_{q^k}$ and therefore n divides $q^k - 1$. That is, for some integer μ, we have $\lambda(\frac{q^k-1}{q-1})\mu = (q - 1)(\frac{q^k-1}{q-1})$, which implies that λ divides $q - 1$. Since we are interested in the construction of codes as the direct sum of two one-weight cyclic codes, then the length n, for such constructed codes, will always be assumed to be given by (1). That is, in general, (1) gives also the length for a constructed two-weight cyclic code, where $\lambda|(q - 1)$.

This work is organized as follows: in Section 2 we recall some already known results about one-weight cyclic codes and two-weight cyclic codes. More specifically, in this section we recall the formula that gives the number of one-weight cyclic codes, and also, we recall the sufficient conditions for the construction of a two-weight cyclic code by means of the direct sum of two one-weight cyclic codes. Section 3 is devoted to present some preliminary results. These results (two lemmas and one corollary) will be fundamental in order to achieve our goal. In Section 4 we present, for some finite fields, a lower bound for the number of two-weight cyclic codes with composite parity-check polynomials. This lower bound will show that there is, indeed, an infinite class of codes for the second case that was mentioned at the beginning of this section. Finally, in Section 5 some examples are shown, whereas Section 6 is devoted to conclusions.

2   Some Already Known Results

First of all, we set, for this section and for the rest of this work, the following:

Notation: By using p, q, m and k, we will denote four positive integers such that p is a prime number and $q = p^m$. For now on, we will fix $\Delta = (q^k - 1)/(q - 1)$, $w_1 = \gcd(\Delta, q - 1)$ and for each integer $\jmath$, with $1 \leq p^\jmath < q$, we set $w_2^{(\jmath)} = \gcd(\Delta, q - 1, p^\jmath - 1)$. Since $\gcd(\Delta, (q - 1)w_2^{(\jmath)}/w_1) = w_2^{(\jmath)}$, we can write w_2 as a linear combination of Δ and $(q - 1)w_2^{(\jmath)}/w_1$ with integer coefficients x_0 and y_0, that is

$$w_2 = \Delta x_0 + \frac{(q - 1)w_2^{(\jmath)}}{w_1} y_0 .$$

Then for each $\jmath$, with $1 \le p^{\jmath} < q$, we will fix the integers x_0 and z_0 in such way that

$$\Delta x_0 \equiv w_2^{(\jmath)} \pmod{\frac{(q-1)w_2^{(\jmath)}}{w_1}} \quad \text{and} \quad z_0 = -x_0 \left(\frac{(p^{\jmath}-1)}{w_2^{(\jmath)}} \right). \tag{2}$$

By using ϕ we will denote the Euler ϕ-function (see, for example, [1, p. 7]), whereas by using δ we will denote the Kronecker's delta ($\delta_x(y)$ is one if $x = y$ and zero otherwise).

Now, we recall some useful definitions:

A linear code is said to be *projective* if the minimum weight of its dual code is at least 3. This means that the columns of a generator matrix are nonzero distinct representatives of the one-dimensional subspaces of $\mathbb{F}_{q^k}$ (the projective points) where k is the dimension of the code.

A linear code is a *N-weight* code if the number of nonzero weights is N.

A cyclic code is *irreducible* if its check polynomial is irreducible (its polynomial representation is a minimal ideal).

For a given length and dimension, it is possible to determine the exact number of one-weight cyclic codes. The following result, that was introduced in [3], gives an explicit formula for such number.

Theorem 1. *Let q, k and Δ be as before. Let γ be a primitive element of $\mathbb{F}_{q^k}$. For a positive integer a, let $h_a(x) \in \mathbb{F}_q[x]$ be the minimal polynomial of γ^a. Let $n = \lambda\Delta$, for some integer $\lambda > 0$, which divides $q - 1$. Then, the following two statements are equivalent:*

A) $\gcd(a, \Delta) = 1$.
B) $h_a(x)$ *is the parity-check polynomial for a one-weight cyclic code of dimension k.*

Additionally, the number of one-weight cyclic codes of length n and dimension k is equal to

$$\delta_\lambda(\gcd(n, q-1)) \frac{\lambda\phi(\Delta)}{k}.$$

Sufficient conditions for the construction of a two-weight cyclic code by means of the direct sum of two one-weight cyclic codes, were recently presented in [4]. The following theorem gives the details of such conditions.

Theorem 2. *Let p, q, k, and Δ be as before. Let γ be a primitive element of $\mathbb{F}_{q^k}$. For a positive integer a, let $h_a(x) \in \mathbb{F}_q[x]$ be the minimal polynomial of γ^a. Also, let a_1, a_2 be integers such that $a_1 q^i \not\equiv a_2 \pmod{q^k - 1}$, for all $i \ge 0$, and where we are assuming that a_2 is a unit in the ring $\mathbb{Z}_\Delta$ (that is $a_2 \in \mathbb{Z}_\Delta^*$), and let $\tilde{a}_2$ be the inverse of a_2 in $\mathbb{Z}_\Delta^*$. In addition, let ν be the integer such that $\nu = \gcd(a_1 - a_2, q - 1)$. For some integer ℓ, such that $\ell | \gcd(a_1, a_2, q - 1)$, we set $\lambda = \frac{(q-1)\ell}{\gcd(a_1, a_2, q-1)}$, $n = \lambda\Delta$ and $\mu = (q - 1)/\lambda$. Suppose that at least one of the following two conditions holds:*

1) $p = 2$, $k = 2$, $\nu = 1$ and a_1 is a unit in the ring $\mathbb{Z}_\Delta$, or
2) for some integer $\jmath$, with $1 \le p^\jmath < q^k$, we have

$$(1 + \tilde{a}_2(a_1 - a_2))p^\jmath \equiv 1 \pmod{\Delta \nu}, \tag{3}$$

then the following four assertions are true:

a) $h_{a_1}(x)$ and $h_{a_2}(x)$ are the parity-check polynomials for two different one-weight cyclic codes of length n and dimension k.
b) $\mu | \nu$ and $\lambda > \nu/\mu$.
c) If C is the cyclic code with parity-check polynomial $h_{a_1}(x) h_{a_2}(x)$, then C is an $[n, 2k]$ two-weight cyclic code with weight enumerator polynomial:

$$A(z) = 1 + (\mu/\nu)n(q-1)z^{(\lambda - (\nu/\mu))q^{k-1}} + (q^{2k} - 1 - (\mu/\nu)n(q-1))z^{\lambda q^{k-1}}.$$

d) C is a projective code if and only if $\nu = \mu$.

Since $\nu = \gcd(a_1 - a_2, q - 1)$ then clearly $\nu | q - 1$, which in turn implies that $\Delta \nu | q^k - 1$ (recall that $\Delta = (q^k - 1)/(q - 1)$), therefore $q^k \equiv 1 \pmod{\Delta \nu}$. Now, since $p^\jmath | q^k$, then there exist an integer $\jmath'$, with $1 \le p^{\jmath'} < q^k$ such that $p^\jmath p^{\jmath'} \equiv 1 \pmod{\Delta \nu}$ (in order to find such $\jmath'$, take $p^{\jmath'} = 1$ if $p^\jmath = 1$ and $p^{\jmath'} = q^k/p^\jmath$ otherwise). Thus, multiplying by $p^{\jmath'}$ in both sides of (3) we get

$$\tilde{a}_2(a_1 - a_2) \equiv p^{\jmath'} - 1 \pmod{\Delta \nu}, \text{ for some integer } \jmath', \text{ with } 1 \le p^{\jmath'} < q^k.$$

If we replace $\jmath'$ by $\jmath$, in previous congruence, we obtain the following:

Remark 1. Since $q^k \equiv 1 \pmod{\Delta \nu}$ then it is important to observe that condition 2), in previous theorem, is equivalent to

$$\tilde{a}_2(a_1 - a_2) \equiv p^\jmath - 1 \pmod{\Delta \nu}, \text{ for some integer } \jmath, \text{ with } 1 \le p^\jmath < q^k.$$

Remark 2. Due to Assertion b), in previous theorem, we conclude that $\lambda > 1$ for all two-weight cyclic codes constructed as a direct sum of two one-weight cyclic codes.

As we will see in Section 4, Theorem 1 and Condition 2), in previous theorem, will be the main tool in order to give a lower bound for the number of two-weight cyclic codes with composite parity-check polynomials.

3   Some Preliminary Results

We begin this section recalling the following:

Fact 1. Let S be a finite set, where a symmetric relation $\sim$ has been established among pairs of elements in S. Suppose that for some fixed integer t and any element x in S, we have that $|\{y \in S : x \ne y \text{ and } x \sim y\}| = t$. Then

$$|\{\{x, y\} \in 2^S : x \ne y \text{ and } x \sim y\}| = \frac{|S|t}{2}.$$

The following is our first result.

Lemma 1. *Let p, q, k, Δ, w_1, $\jmath$, $w_2^{(\jmath)}$ and x_0 be as before. Also, let a_2 and μ be two integers such that $\gcd(\Delta, a_2) = 1$. For each integer i we set $a_1(i) = p^{\jmath}a_2 + i\mu\Delta$ and $\nu(i) = \gcd(a_1(i) - a_2, q - 1)$. Suppose that $\gcd((q-1)/w_1, p^{\jmath} - 1)$ divides $(p^{\jmath}-1)/w_2^{(\jmath)}$. Then there exists an integer t_0 in such way that the following four assertions are true:*

a) t_0 *does not depend on the choice of i.*
b) $p^{\jmath} - 1$ *divides t_0.*
c) $\nu(i) = \gcd(p^{\jmath} - 1 + (t_0 + i\mu)\Delta, \frac{(q-1)w_2^{(\jmath)}}{w_1})$.
d) $\gcd(\Delta, \nu(i)) = w_2^{(\jmath)}$, *for all i.*

Proof. Since $\gcd((q - 1)/w_1, p^{\jmath} - 1)$ divides $(p^{\jmath} - 1)/w_2^{(\jmath)}$, then there exists an integer y_0 such that

$$y_0\left(\frac{(q-1)}{w_1}\right) \equiv -x_0\left(\frac{(a_2 - 1)(p^{\jmath} - 1)}{w_2^{(\jmath)}}\right) \pmod{p^{\jmath} - 1}.$$

Let

$$t_0 = x_0\left(\frac{(a_2 - 1)(p^{\jmath} - 1)}{w_2^{(\jmath)}}\right) + y_0\left(\frac{(q-1)}{w_1}\right). \tag{4}$$

Clearly t_0 does not depend on the choice of i and $p^{\jmath} - 1$ divides t_0.

Now, by hypothesis we have $a_1(i) - a_2 = a_2(p^{\jmath} - 1) + i\mu\Delta$. If $w_1/w_2^{(\jmath)} \neq 1$, then clearly $w_1/w_2^{(\jmath)}$ divides both Δ and $q - 1$, but it does not divide $a_2(p^{\jmath} - 1)$, since $\gcd(\Delta, a_2) = 1$. So, in general, we have

$$\nu(i) = \gcd(a_2(p^{\jmath} - 1) + i\mu\Delta, q - 1)$$
$$= \gcd(a_2(p^{\jmath} - 1) + i\mu\Delta, \frac{(q-1)w_2^{(\jmath)}}{w_1}). \tag{5}$$

On the other hand, since we know that $w_2^{(\jmath)}|\Delta$ then $y_0(q - 1)\Delta/w_1 \equiv 0 \pmod{(q - 1)w_2^{(\jmath)}/w_1}$. Therefore

$$(t_0 + i\mu)\Delta \equiv \Delta x_0\left(\frac{(a_2 - 1)(p^{\jmath} - 1)}{w_2^{(\jmath)}}\right) + i\mu\Delta \pmod{\frac{(q-1)w_2^{(\jmath)}}{w_1}}$$

$$\equiv a_2(p^{\jmath} - 1) - (p^{\jmath} - 1) + i\mu\Delta \pmod{\frac{(q-1)w_2^{(\jmath)}}{w_1}},$$

since $\Delta x_0 \equiv w_2^{(\jmath)} \pmod{(q - 1)w_2^{(\jmath)}/w_1}$. Thus, by using (5), the third assertion follows.

Let $d = \gcd(\Delta, \nu(i))$. Since $w_2^{(\jmath)} = \gcd(\Delta, q - 1, p^{\jmath} - 1)$ and $\nu(i) = \gcd(a_2(p^{\jmath} - 1) + i\mu\Delta, q - 1)$, then $w_2^{(\jmath)}|\nu(i)$ and therefore $w_2^{(\jmath)}|d$. Now, if $d|\Delta$ and $d|\nu(i)$, then by using the third assertion, and since $\nu(i) = \gcd(a_1(i) - a_2, q - 1)$, we have that $d|q - 1$ and $d|p^{\jmath} - 1$, thus $d|w_2^{(\jmath)}$. Therefore $d = w_2^{(\jmath)}$. □

By taking advantage of previous result, we now present the following:

Lemma 2. *Let p, q, k, Δ, w_1, $\jmath$, $w_2^{(\jmath)}$ and x_0 be as before. Let λ and μ be two integers. Also, let a_2 be an integer such that a_2 is a unit in the ring $\mathbb{Z}_\Delta$. Let $\tilde{a}_2$ be the inverse of a_2 in $\mathbb{Z}_\Delta^*$. Let t_0 be the integer that one obtains under Lemma 1. Let $I = \{0, 1, 2, \ldots, \lambda - 1\}$ and $T = \{t_0, t_0 + \mu, t_0 + 2\mu, \ldots, t_0 + (\lambda - 1)\mu\}$. For each $i \in I$, we set $a_1(i) = p^\jmath a_2 + i\mu\Delta$ and $\nu(i) = \gcd(a_1(i) - a_2, q - 1)$. Suppose that $\gcd((q-1)/w_1, p^\jmath - 1)$ divides $(p^\jmath - 1)/w_2^{(\jmath)}$. Then there exists a bijective map $\mathcal{B} : I \to T$ in such way that, for each $i \in I$, the following two assertions are true:*

a) *If $\tilde{a}_2(a_1(i) - a_2) \equiv p^\jmath - 1 \pmod{\Delta\nu(i)}$ or $\nu(i)|\mathcal{B}(i)$, then $\nu(i)|(p^\jmath - 1)$.*
b) *$\tilde{a}_2(a_1(i) - a_2) \equiv p^\jmath - 1 \pmod{\Delta\nu(i)} \iff \nu(i)|\mathcal{B}(i)$.*

Proof. For each $i \in I$, we take the uniquely determined integer $\mathcal{B}(i) \in T$ such that $\mathcal{B}(i) = t_0 + i\mu$. Due to Assertion a), in previous lemma, $\mathcal{B}$ is clearly a bijective map.

Since $a_2\tilde{a}_2 \equiv 1 \pmod{\Delta}$, then there exists an integer ℓ such that $a_2\tilde{a}_2 = 1 + \ell\Delta$, but we know that $a_1(i) - a_2 = a_2(p^\jmath - 1) + i\mu\Delta$, thus

$$\tilde{a}_2(a_1(i) - a_2) = p^\jmath - 1 + (\ell(p^\jmath - 1) + \tilde{a}_2 i\mu)\Delta. \tag{6}$$

On the other hand, by Assertion d) in previous lemma, we have $w_2^{(\jmath)}|\nu(i)$. By keeping this in mind, let d be the greatest integer satisfying:

- $d|\nu(i)$ and
- $\pi|d \iff \pi|w_2^{(\jmath)}$, for all prime numbers π.

Clearly $w_2^{(\jmath)}|d$ and $\gcd(d, \nu(i)/d) = 1$.

Now, we are going to prove the first assertion. So, if $\tilde{a}_2(a_1(i) - a_2) \equiv p^\jmath - 1 \pmod{\Delta\nu(i)}$, thus $\nu(i)|(\tilde{a}_2(a_1(i) - a_2)) - (p^\jmath - 1)$. But $\nu(i)|(a_1(i) - a_2)$ therefore $\nu(i)|(p^\jmath - 1)$. On the other hand, if $\nu(i)|\mathcal{B}(i)$, then $\nu(i)|(t_0 + i\mu)$. Thus, by virtue of Assertion c) in previous lemma, we have $\nu(i)|(p^\jmath - 1)$.

Now, we are going to prove the second assertion. So, if $\tilde{a}_2(a_1(i) - a_2) \equiv p^\jmath - 1 \pmod{\Delta\nu(i)}$, then, by using (6), we conclude $\nu(i)|(\ell(p^\jmath - 1) + \tilde{a}_2 i\mu)$, but, due to Assertion a), we know that $\nu(i)|(p^\jmath - 1)$, therefore this implies that $\nu(i)|\tilde{a}_2 i\mu$. Now $\gcd(\Delta, \nu(i)) = w_2^{(\jmath)}$ and $\tilde{a}_2 \in \mathbb{Z}_\Delta^*$, thus $d|i\mu$. On the other hand, since $\nu(i)|(p^\jmath - 1)$ then, by virtue of Assertion c) in previous lemma, we have that $\nu(i)|(t_0 + i\mu)\Delta$. Again $\gcd(\Delta, \nu(i)) = w_2^{(\jmath)}$, thus $(\nu(i)/d)|(t_0 + i\mu)$. But, by virtue of Assertion b) in previous lemma, and since $(\nu(i)/d)|(p^\jmath - 1)$, we have $(\nu(i)/d)|t_0$, then $(\nu(i)/d)|i\mu$. In this way, what we have proven is $\gcd(d, \nu(i)/d) = 1$, $d|i\mu$ and $(\nu(i)/d)|i\mu$, therefore $\nu(i)|i\mu$. Now Assertion b), in previous lemma, implies that $\nu(i)|t_0$, and since $\nu(i)|i\mu$, then the final conclusion is $\nu(i)|\mathcal{B}(i)$.

Conversely, suppose that $\nu(i)|\mathcal{B}(i)$. Again, due to Assertion a), we know that $\nu(i)|(p^\jmath - 1)$. Then, by using Assertion b) in previous lemma, we have $\nu(i)|t_0$, so $\nu(i)|i\mu$, since $\mathcal{B}(i) = t_0 + i\mu$. Thus what we have is $\nu(i)|(p^\jmath - 1)$ and $\nu(i)|i\mu$, therefore $\ell(p^\jmath - 1) + \tilde{a}_2 i\mu \equiv 0 \pmod{\nu(i)}$, for any integer ℓ. Therefore, by using (6), we have the result. $\square$

We end this section with the following:

Corollary 1. *Let p, q, k, Δ, w_1, $\jmath$, $w_2^{(\jmath)}$, x_0 and z_0 be as before. Let λ and μ be two integers such that $\lambda\mu = q - 1$. Let $I = \{0, 1, 2, \ldots, \lambda - 1\}$ and $Z = \{z_0, z_0 + \mu, z_0 + 2\mu, \ldots, z_0 + (\lambda - 1)\mu\}$. Let ν_0 be any positive integer such that $\nu_0 | (q - 1)$. For each choice of $\jmath$ and ν_0 we set the integer $N_{\nu_0}^{(\jmath)}$ given by*

$$N_{\nu_0}^{(\jmath)} = |\{z \in Z : \gcd(p^{\jmath} - 1 + z\Delta, \frac{(q-1)w_2^{(\jmath)}}{w_1}) = \nu_0 \text{ and } \nu_0|z\}| . \quad (7)$$

Also, let a_2 be an integer such that $\mu | a_2$ and a_2 is a unit in the ring $\mathbb{Z}_\Delta$. Let $\tilde{a}_2$ be the inverse of a_2 in $\mathbb{Z}_\Delta^*$. For each $i \in I$, we set $a_1(i) = p^{\jmath}a_2 + i\mu\Delta$ and $\nu(i) = \gcd(a_1(i) - a_2, q - 1)$. Suppose that $\gcd((q-1)/w_1, p^{\jmath} - 1)$ divides $(p^{\jmath} - 1)/w_2^{(\jmath)}$. Then

$$|\{i \in I : \tilde{a}_2(a_1(i) - a_2) \equiv p^{\jmath} - 1 \pmod{\Delta\nu(i)} \text{ and } \nu(i) = \nu_0\}| = N_{\nu_0}^{(\jmath)} . \quad (8)$$

Proof. Let t_0 be the integer that one obtains under Lemma 1. Also, let $T = \{t_0, t_0 + \mu, t_0 + 2\mu, \ldots, t_0 + (\lambda - 1)\mu\}$. By using (2) and (4) we have

$$t_0 = z_0 + x_0 \left(\frac{a_2(p^{\jmath} - 1)}{w_2^{(\jmath)}} \right) + y_0 \left(\frac{(q-1)}{w_1} \right) .$$

Since $\mu | a_2$ and $a_2 \in \mathbb{Z}_\Delta^*$, then $\mu \in \mathbb{Z}_\Delta^*$. But $\gcd(\Delta, q-1) = w_1$ thus $\mu | ((q-1)/w_1)$. In consequence we have that μ divides both $x_0 a_2(p^{\jmath}-1)/w_2^{(\jmath)}$ and $y_0(q-1)/w_1$, therefore, and since $\lambda\mu = q-1$, we have that for each $i \in I$, there exists a uniquely determined integer $z \in Z$ such that $z \equiv t_0 + i\mu \pmod{q-1}$. Since $\nu_0 | (q-1)$ then $\nu_0 | z$ if and only if $\nu_0 | (t_0 + i\mu)$. Thus, as a direct consequence of the fact that $\mathcal{B} : I \to T$, in previous lemma, is a bijective map, and due to Assertions c) and b) in Lemmas 1 and 2, respectively, we can now conclude that for each $i \in I$, there exists a uniquely determined integer $z \in Z$ such that

$$\tilde{a}_2(a_1(i) - a_2) \equiv p^{\jmath} - 1 \pmod{\Delta\nu(i)} \text{ and } \nu(i) = \nu_0$$

if and only if

$$\gcd(p^{\jmath} - 1 + (t_0 + i\mu)\Delta, \frac{(q-1)w_2^{(\jmath)}}{w_1}) = \nu_0 \text{ and } \nu_0|(t_0 + i\mu)$$

if and only if

$$\gcd(p^{\jmath} - 1 + z\Delta, \frac{(q-1)w_2^{(\jmath)}}{w_1}) = \nu_0 \text{ and } \nu_0|z ,$$

since we know that $z \equiv t_0 + i\mu \pmod{q-1}$, $((q-1)w_2^{(\jmath)}/w_1)|(q^k - 1)$ and $\nu_0|(t_0 + i\mu) \Leftrightarrow \nu_0|z$. $\qquad \square$

Remark 3. It is important to emphasize that the value of the integer $N_{\nu_0}^{(\jmath)}$, in (7), does not depend on the choice of a_2, therefore equality (8) will hold for all unit a_2 in the ring $\mathbb{Z}_\Delta$.

4   A Lower Bound

Let p, q, k and Δ be as before. Let γ be a primitive element of $\mathbb{F}_{q^k}$. For a positive integer a, let $h_a(x) \in \mathbb{F}_q[x]$ be the minimal polynomial of γ^a. Let n be an integer such that $n = \lambda\Delta$, for some positive integer λ which divides $q - 1$. Let μ be the integer such that $\lambda\mu = q - 1$ and for now on, we are going to suppose that μ is a unit in the ring $\mathbb{Z}_\Delta$. Let $\tilde{\mu}$ be the inverse of μ in $\mathbb{Z}_\Delta^*$. If b is an integer, with $0 \le b < n$, then we define the *cyclotomic coset modulo n* over $\mathbb{F}_q$ which contains b as (see for example [2, p. 197]):

$$C_b = \{b,\ bq \ (\mathrm{mod}\ n),\ bq^2 \ (\mathrm{mod}\ n),\ \ldots,\ bq^{m_b-1} \ (\mathrm{mod}\ n)\}\,,$$

where x (mod n) is the remainder of x divided by n, and m_b is the smallest positive integer such that $bq^{m_b} \equiv b \pmod{n}$. The subscript b is called the *coset representative modulo n*.

Since $\mu n = q^k - 1$, then the set $U = \{\gamma^{\mu b} : 0 \le b < n\}$ contains all the zeros of $x^n - 1$ and therefore $h_{\mu b}(x) | (x^n - 1)$ for all integer b with $0 \le b < n$. That is $h_{\mu b}(x)$ is the parity-check polynomial for an irreducible cyclic code of length n. Taking into consideration all of this, we define the following set of cyclotomic cosets modulo n:

$$S = \{C_b : C_b \text{ is cyclotomic coset modulo } n \text{ over } \mathbb{F}_q \text{ and } \gcd(b, \Delta) = 1\}\,.$$

Observe that $\gcd(q, \Delta) = 1$ and $\Delta|n$, therefore the previous set is well defined since it does not depend on the choice of the coset representative modulo n, b, in each cyclotomic coset modulo n over $\mathbb{F}_q$, C_b. Now we define the function $\mathcal{P} : S \to \mathbb{F}_q[x]$, in such a way that $\mathcal{P}(C_b) = h_{\mu b}(x)$. Since $\gcd(\mu b, \Delta) = 1$, for each $C_b \in S$, then, in light of Theorem 1, we can see that $h_{\mu b}(x)$ is the parity-check polynomial for a one-weight cyclic code of length n and dimension k, therefore $|C_b| = k$ for all cyclotomic coset C_b in S. Additionally, the number of elements in S corresponds to the number of one-weight cyclic codes of length n and dimension k, that is

$$|S| = |Im(\mathcal{P})| = \delta_\lambda(\gcd(n, q - 1))\frac{\lambda\phi(\Delta)}{k}\,. \tag{9}$$

We want to establish a lower bound for the number of two-weight cyclic codes that are constructed by means of the direct sum of two one-weight cyclic codes of length n and dimension k. That is, we want to establish a lower bound for the cardinality of the following set

$$\{\{C_{b_1}, C_{b_2}\} \in 2^S : C_{b_1} \ne C_{b_2} \text{ and } h_{\mu b_1}(x)h_{\mu b_2}(x) \text{ is the parity-check}$$
$$\text{polynomial for a two-weight cyclic code}\}\,.$$

In order to establish such lower bound, the idea now is to take advantage of Fact 1. In doing this it is important, first, to define a symmetric relation among pairs of elements in S. The following definition and lemma take care of this.

Definition 1. *Let p, q, k and Δ be as before. Let μ, $\tilde{\mu}$ and S be as in the previous discussion. For each pair of cyclotomic cosets C_{b_1} and C_{b_2} in S, we will say that C_{b_1} is related to C_{b_2}, denoted $C_{b_1} \sim C_{b_2}$, if and only if the coset representatives b_1 and b_2, in this order, satisfy*

$$\tilde{\mu}\tilde{b}_2(\mu b_1 - \mu b_2) \equiv p^{\jmath} - 1 \pmod{\Delta\nu}, \text{ for some integer } \jmath, \text{ with } 1 \le p^{\jmath} < q^k,$$
(10)

where $\tilde{b}_2$ is the inverse of b_2 in $\mathbb{Z}_\Delta^*$ and $\nu = \gcd(\mu b_1 - \mu b_2, q-1)$.

Since any element in a cyclotomic coset could be the coset representative, then it is rather clear that it is necessary to prove that the previous relation is well defined. The following lemma shows that "$\sim$" is, indeed, well defined.

Lemma 3. *Let p, q, k and Δ be as before. Let n, λ, μ, $\tilde{\mu}$ and S be as in previous discussion. Let C_{b_1} and C_{b_2} be in S, and let $\tilde{b}_1$ and $\tilde{b}_2$ be the inverses of b_1 and b_2, respectively, in $\mathbb{Z}_\Delta^*$. Also let $\nu = \gcd(\mu b_1 - \mu b_2, q-1)$. Suppose that the coset representatives b_1 and b_2 satisfy, in this order, (10). Then the following three assertions are true:*

a) *b_2 and b_1 satisfy, in this order, (10). That is, for some integer $\jmath'$, with $1 \le p^{\jmath'} < q^k$, we have*

$$\tilde{\mu}\tilde{b}_1(\mu b_2 - \mu b_1) \equiv p^{\jmath'} - 1 \pmod{\Delta\nu}.$$

b) *Let r be a positive integer and ℓ any integer. If $\nu' = \gcd(\mu(b_1 q^r + \ell n) - \mu b_2, q-1)$ then $\nu' = \nu$, and $b_1 q^r + \ell n$ and b_2 satisfy, in this order, (10). That is, for some integer $\jmath'$, with $1 \le p^{\jmath'} < q^k$, we have*

$$\tilde{\mu}\tilde{b}_2(\mu(b_1 q^r + \ell n) - \mu b_2) \equiv p^{\jmath'} - 1 \pmod{\Delta\nu}.$$

In addition, the integer $\jmath'$ is uniquely determined by $p^{\jmath'} \equiv p^{\jmath}q^r \pmod{\Delta\nu}$.

c) *The relation $\sim$, in previous definition, is a well defined reflexive and symmetric relation.*

Proof. If b_1 and b_2 satisfy (10), then $\tilde{b}_2 b_1 \equiv p^{\jmath} \pmod{\Delta}$, which implies that $\tilde{b}_2 \equiv \tilde{b}_1 p^{\jmath} \pmod{\Delta}$. Therefore $\tilde{b}_2 = \tilde{b}_1 p^{\jmath} + \ell\Delta$ for some integer ℓ. By substituting $\tilde{b}_2$ into (10), and since $\nu | (\mu b_1 - \mu b_2)$, we have $\tilde{\mu}\tilde{b}_1(\mu b_2 - \mu b_1)p^{\jmath} \equiv 1 - p^{\jmath} \pmod{\Delta\nu}$. Since $q^k \equiv 1 \pmod{\Delta\nu}$, then let $\jmath'$, with $1 \le p^{\jmath'} < q^k$, such that $p^{\jmath}p^{\jmath'} \equiv 1 \pmod{\Delta\nu}$. By multiply both sides of $\tilde{\mu}\tilde{b}_1(\mu b_2 - \mu b_1)p^{\jmath} \equiv 1 - p^{\jmath} \pmod{\Delta\nu}$ by $p^{\jmath'}$, we have Assertion a).

Clearly $\nu' = \gcd(\mu b_1 - \mu b_2 + \mu b_1(q^r - 1) + \mu\ell n, q-1) = \nu$, since $\mu n = q^k - 1$ and since $(q-1)$ divides both $(q^k - 1)$ and $(q^r - 1)$. Now, for the second part, observe first that $\tilde{\mu}\mu\tilde{b}_2 b_2 = 1 + \ell'\Delta$ for some integer ℓ'. Thus $\tilde{\mu}\mu\tilde{b}_2 b_1 = 1 + \tilde{b}_2(\mu b_1 - \mu b_2) + \ell'\Delta$. Therefore

$$\begin{aligned}
1 + \tilde{\mu}\tilde{b}_2(\mu(b_1 q^r + \ell n) - \mu b_2) &= \tilde{\mu}\mu\tilde{b}_2 b_1 q^r + \tilde{\mu}\mu\tilde{b}_2\ell n - \ell'\Delta \\
&= (1 + \tilde{\mu}\tilde{b}_2(\mu b_1 - \mu b_2) + \ell'\Delta)q^r + \tilde{\mu}\mu\tilde{b}_2\ell n - \ell'\Delta \\
&= (1 + \tilde{\mu}\tilde{b}_2(\mu b_1 - \mu b_2))q^r + \tilde{\mu}\mu\tilde{b}_2\ell n + \ell'\Delta(q^r - 1).
\end{aligned}$$

Since $q^k \equiv 1 \pmod{\Delta\nu}$, then let $\jmath'$ be the uniquely determined integer such that $1 \le p^{\jmath'} < q^k$ and $p^{\jmath'} \equiv p^{\jmath}q^r \pmod{\Delta\nu}$. Now $\nu|(q^r - 1)$ and $\Delta\nu|\mu n$, thus the conclusion is

$$1 + \tilde{\mu}\tilde{b}_2(\mu(b_1q^r + \ell n) - \mu b_2) \equiv (1 + \tilde{\mu}\tilde{b}_2(\mu b_1 q - \mu b_2))q^r \pmod{\Delta\nu}$$
$$\equiv p^{\jmath}q^r \equiv p^{\jmath'} \pmod{\Delta\nu} ,$$

which in turn gives the proof of Assertion b).

Since any element b' in a cyclotomic coset modulo n, C_b, is in the form $b' = bq^r + \ell n$, for some integer ℓ, then a recursive application of Assertions a) and b) shows that relation $\sim$ is well defined. The reflexive property is immediate for (10) (taking $p^{\jmath} = 1$), whereas the symmetric property was already proved by Assertion a). □

Let $B = \{0, 1, 2, \ldots, n-1\}$ and $I = \{0, 1, 2, \ldots, \lambda-1\}$. Now, we are going to take a fixed b_2 in set B, in such way that $\gcd(\Delta, b_2) = 1$. Let $\tilde{b}_2$, μ and $\tilde{\mu}$ be as before, and set $a_2 = \mu b_2$ and $\tilde{a}_2 = \tilde{\mu}\tilde{b}_2$. Also, for each $i \in I$, set $a_1(i) = p^{\jmath}a_2 + i\mu\Delta$ and $\nu(i) = \gcd(a_1(i) - a_2, q - 1)$. Let w_1, $\jmath$, $w_2^{(\jmath)}$ be as before. Now, for each pair of integers, $\jmath$ and ν_0, satisfying: $1 \le p^{\jmath} < q$, $\nu_0|(q-1)$ and $\gcd((q-1)/w_1, p^{\jmath} - 1)$ divides $(p^{\jmath} - 1)/w_2^{(\jmath)}$, we define

$$B_{\nu_0}^{(\jmath)} = \{b \in B : \tilde{\mu}\tilde{b}_2(\mu b - \mu b_2) \equiv p^{\jmath} - 1 \pmod{\Delta\nu_0} \text{ and } \gcd(\mu b - \mu b_2, q-1) = \nu_0\}.$$

We are interested in obtaining the number of elements of $B_{\nu_0}^{(\jmath)}$. However, before this, observe that $C_{b_2} \in S$ and if $b_1 \in B_{\nu_0}^{(\jmath)}$, then $b_1 \equiv p^{\jmath}b_2 \pmod{\Delta}$, so $\gcd(b_1, \Delta) = 1$, since $\gcd(p^{\jmath}, \Delta) = 1$. Therefore C_{b_1} is also in S. On the other hand, if $b_1 \equiv p^{\jmath}b_2 \pmod{\Delta}$ then, since $n = \lambda\Delta$, there must exists a uniquely determined integer $i \in I$ such that $b_1 \equiv p^{\jmath}b_2 + i\Delta \pmod{n}$, which implies that $p^{\jmath}a_2 + i\mu\Delta = a_1(i) \equiv \mu b_1 \pmod{q^k - 1}$, since $\mu n = q^k - 1$. That is, $a_1(i) = \mu b_1 + \ell(q^k - 1)$, for some integer ℓ. But clearly $(q-1)|(q^k - 1)$, thus $\gcd(\mu b_1 - \mu b_2, q-1) = \nu(i)$. Additionally, since $\Delta\nu_0|(q^k - 1)$, then $b_1 \in B_{\nu_0}^{(\jmath)}$ if and only if $\tilde{a}_2(a_1(i) - a_2) \equiv p^{\jmath} - 1 \pmod{\Delta\nu(i)}$. Therefore the main conclusion is that for each $b_1 \in B$, such that $b_1 \equiv p^{\jmath}b_2 \pmod{\Delta}$, there exists a uniquely determined integer $i \in I$, in such way that $b_1 \in B_{\nu_0}^{(\jmath)}$ if and only if $\tilde{a}_2(a_1(i) - a_2) \equiv p^{\jmath} - 1 \pmod{\Delta\nu(i)}$ and $\nu(i) = \nu_0$. Considering this, and equality (8) in Corollary 1, we can now say that

$$|B_{\nu_0}^{(\jmath)}| = N_{\nu_0}^{(\jmath)} = N_{\mu d}^{(\jmath)} , \qquad (11)$$

where $d = \nu_0/\mu$. Observe that integer d always divides λ.

Remark 4. By using the same notation, observe that if $C_{b_1} \sim C_{b_2}$ and if $d = 1$ (that is, if $\gcd(\mu b_1 - \mu b_2, q-1) = \mu$), then $C_{b_1} \ne C_{b_2}$, because if $C_{b_1} = C_{b_2}$, then $\nu_0 = q - 1$, which in turn implies that $d = \lambda > 1$, due to Remark 2.

Furthermore, we can still say something else about the sets $B_{\nu_0}^{(\jmath)}$:

Lemma 4. *Let p, q, m and k be as before. By using the same notation and hypothesis as in the previous discussion, let $C_{b_1} \in S$ such that $C_{b_1} \sim C_{b_2}$. Since $C_{b_1} \sim C_{b_2}$, then let $\nu_0 = \gcd(\mu b_1 - \mu b_2, q - 1)$ and $\jmath$, with $1 \le p^{\jmath} < q^k$, such that $\tilde{\mu}\tilde{b}_2(\mu b_1 - \mu b_2) \equiv p^{\jmath} - 1 \pmod{\Delta\nu_0}$. Then there exist uniquely determined integers b_1' and $\jmath'$ such that $b_1' \in C_{b_1}$, $1 \le p^{\jmath'} < q$ and $b_1' \in B_{\nu_0}^{(\jmath')}$ (and therefore $C_{b_1'} \sim C_{b_2}$).*

Proof. By using the division algorithm, let s and $\jmath'$ be the two uniquely determined integers such that $\jmath = ms + \jmath'$, with $0 \le \jmath' < m$. On the other hand, let r be the uniquely determined integer such that $r + s \equiv 0 \pmod{k}$, with $0 \le r < k$. Now, let b_1' be the uniquely determined element in C_{b_1} such that $b_1' \equiv b_1 q^r \pmod{n}$. Since $q = p^m$ and $q^k \equiv 1 \pmod{\Delta\nu_0}$, then clearly $p^{\jmath'} \equiv p^{\jmath} q^r \pmod{\Delta\nu_0}$. Therefore, by means of Assertion b) in previous lemma, we conclude that $b_1' \in B_{\nu_0}^{(\jmath')}$. Finally by using the second part of the same assertion, we can see that the integers b_1' and $\jmath'$ do not depend on the choice of the coset representative in C_{b_1}. $\square$

We observe from above that for each $\jmath'$, with $0 \le \jmath' < m$, there are exactly k different choices for pairs $(\jmath, r)$, with $1 \le p^{\jmath}, q^r < q^k$, such that $p^{\jmath'} \equiv p^{\jmath} q^r \pmod{\Delta\nu_0}$. Therefore, if d is an integer such that $d|\lambda$, then, by virtue of the previous lemma and equality (11), we have

$$|\{b_1 \in B : C_{b_1} \sim C_{b_2} \text{ and } \gcd(\mu b_1 - \mu b_2, q - 1) = \mu d\}| = k \sum_{\jmath=0}^{m-1} N_{\mu d}^{(\jmath)}.$$

But we already said that all the cyclotomic cosets modulo n, in S, have exactly k elements, thus

$$|\{C_{b_1} \in S : C_{b_1} \sim C_{b_2} \text{ and } \gcd(\mu b_1 - \mu b_2, q - 1) = \mu d\}| = \sum_{\jmath=0}^{m-1} N_{\mu d}^{(\jmath)}.$$

The previous equality and the fact that $\sim$ is reflexive, give us the proof of the following:

Corollary 2. *Let p, q, m and k be as before. Assume the same notation and hypothesis as in the previous lemma and discussion. If C_{b_2} is in S, then*

$$|\{C_{b_1} \in S : C_{b_1} \ne C_{b_2} \text{ and } C_{b_1} \sim C_{b_2}\}| = \sum_{\jmath=0}^{m-1} \sum_{d|\lambda} N_{\mu d}^{(\jmath)} - 1.$$

Clearly, the "-1" term, in the previous equality, is due to the fact that we have to subtract the reflexive instance: $C_{b_2} \sim C_{b_2}$.

The following result establish, for some finite fields, a lower bound for the number of two-weight cyclic codes with composite parity-check polynomials.

Theorem 3. *Let p, q, m, k, Δ, w_1, $\jmath$, $w_2^{(\jmath)}$ be as before. Let μ, λ and n be integers such that $\lambda > 1$, $\mu\lambda = q - 1$ and $n = \lambda\Delta$. Assume that the finite field $\mathbb{F}_q$ and the extension field $\mathbb{F}_{q^k}$ are in such a way that $\gcd((q-1)/w_1, p^\jmath - 1)$ divides $(p^\jmath - 1)/w_2^{(\jmath)}$ for each $\jmath$, with $0 \le \jmath < m$. Then the number of two-weight cyclic codes constructed as the direct sum of two different one-weight cyclic codes of length n and dimension k is at least*

$$\delta_\lambda(\gcd(n, q-1))\frac{\lambda\phi(\Delta)}{2k}\left(\sum_{\jmath=0}^{m-1}\sum_{d|\lambda}N_{\mu d}^{(\jmath)} - 1\right).$$

Additionally, the number of two-weight projective cyclic codes constructed as the direct sum of two different one-weight cyclic codes of length n and dimension k is at least

$$\delta_\lambda(\gcd(n, q-1))\frac{\lambda\phi(\Delta)}{2k}\sum_{\jmath=0}^{m-1}N_\mu^{(\jmath)}.$$

Proof. The first part is a direct consequence of Theorem 2, Remark 1, Fact 1, equality (9) and the previous corollary. For the projective part we also use the Assertion d) of Theorem 2 and Remark 4 (in this context, Remark 4 says that we do not have to subtract the reflexive instance for this part). $\qquad\square$

5   Some Examples

In this section we present some examples of the lower bound that was introduced in Theorem 3.

Let $p = 3$, $q = 9$ and $k = 3$, thus $m = 2$, $\Delta = 91$, $\phi(\Delta) = \phi(13) \cdot \phi(7) = 72$ and $w_1 = w_2^{(\jmath)} = 1$, for $\jmath = 0, 1$. Clearly, the finite field $\mathbb{F}_9$ and the extension field $\mathbb{F}_{729}$ satisfy that $\gcd((q-1)/w_1, p^\jmath - 1)$ divides $(p^\jmath - 1)/w_2^{(\jmath)}$ for each $\jmath$, with $0 \le \jmath < 2$.

If $n = 182$, then $\lambda = 2$ and $\mu = 4$. Now, if $\jmath = 0$ then, by using (2), we have $x_0 = 3$, $z_0 = 0$ and $Z = \{z_0, z_0 + \mu, z_0 + 2\mu, \ldots, z_0 + (\lambda - 1)\mu\} = \{0, 4\}$, therefore, by means of (7), we obtain $N_4^{(0)} = N_8^{(0)} = 1$. On the other hand, if $\jmath = 1$ then, $z_0 = -6$ and $Z = \{-6, -2\}$, therefore $N_4^{(1)} = N_8^{(1)} = 0$. By using Theorem 3 the conclusion is that there are at least 24 two-weight cyclic codes, over $\mathbb{F}_9$, constructed as a direct sum of two different one-weight cyclic codes of length 182 and dimension 3. Additionally, all these 24 two-weight cyclic codes are projective. By means of an exhaustive search, with the help of a computer program, we can see that 24 is exactly the number of two-weight cyclic codes, over $\mathbb{F}_9$, constructed as a direct sum of two different one-weight cyclic codes of length 182 and dimension 3, and also, we can see that all of them are projective.

If $n = 364$, then $\lambda = 4$ and $\mu = 2$. Now, if $\jmath = 0$ then, by using (2), we have $x_0 = 3$, $z_0 = 0$ and $Z = \{z_0, z_0 + \mu, z_0 + 2\mu, \ldots, z_0 + (\lambda - 1)\mu\} = \{0, 2, 4, 6\}$, therefore, by means of (7), we obtain $N_4^{(0)} = N_8^{(0)} = 1$ and $N_2^{(0)} = 2$. On the other hand, if $j = 1$ then, $z_0 = -6$ and $Z = \{-6, -4, -2, 0\}$, therefore $N_4^{(1)} = N_8^{(1)} = 0$ and $N_2^{(1)} = 2$. By using Theorem 3 the conclusion is that there are at least 240 two-weight cyclic codes, over $\mathbb{F}_9$, constructed as a direct sum of two different one-weight cyclic codes of length 364 and dimension 3. Additionally, 192 of these 240 are projective codes. By means of an exhaustive search we can see that 240 is exactly the number of two-weight cyclic codes, over $\mathbb{F}_9$, constructed as a direct sum of two different one-weight cyclic codes of length 364 and dimension 3, and also, we can see that 192 of these 240 are projective codes.

If $n = 728$, then $\lambda = 8$ and $\mu = 1$. Now, if $j = 0$ then, by using (2), we have $x_0 = 3$, $z_0 = 0$ and $Z = \{z_0, z_0 + \mu, z_0 + 2\mu, \ldots, z_0 + (\lambda - 1)\mu\} = \{0, 1, 2, 3, 4, 5, 6, 7\}$, therefore, by means of (7), we obtain $N_4^{(0)} = N_8^{(0)} = 1$, $N_2^{(0)} = 2$ and $N_1^{(0)} = 4$. On the other hand, if $j = 1$ then, $z_0 = -6$ and $Z = \{-6, -5, -4, -3, -2, -1, 0, 1\}$, therefore $N_4^{(1)} = N_8^{(1)} = 0$, $N_2^{(1)} = 2$ and $N_1^{(1)} = 4$. By using Theorem 3 the conclusion is that there are at least 1248 two-weight cyclic codes, over $\mathbb{F}_9$, constructed as a direct sum of two different one-weight cyclic codes of length 728 and dimension 3. Additionally, 768 of these 1248 are projective codes. By means of an exhaustive search we can see that 1248 is exactly the number of two-weight cyclic codes, over $\mathbb{F}_9$, constructed as a direct sum of two different one-weight cyclic codes of length 728 and dimension 3, and also, we can see that 768 of these 1248 are projective codes.

6   Conclusion

We have established, for some finite fields, a lower bound for the number of two-weight (projective or not projective) cyclic codes constructed as direct sum of two different one-weight cyclic codes of the same length and dimension. This lower bound show that there is, indeed, an infinite class of two-weight cyclic codes constructed as direct sum of two different one-weight cyclic codes. For some examples, we saw that such lower bound coincide with the exact number of such codes. In fact, we believe that such coincidence, in these examples, is something more than a fortunate choice.

References

1. Lidl, R., Niederreiter, H.: Finite Fields. Cambridge Univ. Press, Cambridge (1983)
2. MacWilliams, F.J., Sloane, N.J.A.: The Theory of Error-Correcting Codes, Amsterdam. North-Holland, The Netherlands (1977)
3. Vega, G.: Determining the Number of One-weight Cyclic Codes when Length and Dimension are Given. In: Carlet, C., Sunar, B. (eds.) WAIFI 2007. LNCS, vol. 4547, pp. 284–293. Springer, Heidelberg (2007)
4. Vega, G.: Two-weight cyclic codes constructed as the direct sum of two one-weight cyclic codes, Finite Fields Appl. (in press, 2008), doi:10.1016/j.ffa.2008.01.002
5. Wolfmann, J.: Are 2-Weight Projective Cyclic Codes Irreducible? IEEE Trans. Inform. Theory. 51, 733–737 (2005)

On Field Size and Success Probability in Network Coding

Olav Geil[1], Ryutaroh Matsumoto[2], and Casper Thomsen[1]

[1] Department of Mathematical Sciences, Aalborg University, Denmark
olav@math.aau.dk,
caspert@math.aau.dk
[2] Department of Communications and Integrated Systems,
Tokyo Institute of Technology, Japan
ryutaroh@rmatsumoto.org

Abstract. Using tools from algebraic geometry and Gröbner basis theory we solve two problems in network coding. First we present a method to determine the smallest field size for which linear network coding is feasible. Second we derive improved estimates on the success probability of random linear network coding. These estimates take into account which monomials occur in the support of the determinant of the product of Edmonds matrices. Therefore we finally investigate which monomials can occur in the determinant of the Edmonds matrix.

Keywords: Distributed networking, linear network coding, multicast, network coding, random network coding.

1   Introduction

In a traditional data network, an intermediate node only forwards data and never modifies them. Ahlswede et al. [1] showed that if we allow intermediate nodes to process their incoming data and output modified versions of them then maximum throughput can increase, and they also showed that the maximum throughput is given by the minimum of maxflows between the source node and a sink node for single source multicast on an acyclic directional network. Such processing is called network coding. Li et al. [10] showed that computation of linear combinations over a finite field by intermediate nodes is enough for achieving the maximum throughput. Network coding only involving linear combinations is called linear network coding. The acyclic assumption was later removed by Koetter and Médard [9].

In this paper we shall concentrate on the error-free, delay-free multisource multicast network connection problem where the sources are uncorrelated. However, the proposed methods described can be generalized to deal with delays as in [7]. The only exception is the description in Section 7.

Considering multicast, it is important to decide whether or not all receivers (called sinks) can recover all the transmitted information from the senders (called

J. von zur Gathen, J.L. Imaña, and Ç.K. Koç (Eds.): WAIFI 2008, LNCS 5130, pp. 157–173, 2008.

sources). It is also important to decide the minimum size q of the finite field $\mathbf{F}_q$ required for linear network coding.

Before using linear network coding we have to decide coefficients in linear combinations computed by intermediate nodes. When the size q of a finite field is large, it is shown that random choice of coefficients allows all sinks to recover the original transmitted information with high probability [7]. Such a method is called random linear network coding and the probability is called success probability. As to random linear network coding the estimation or determination of the success probability is very important. Ho et al. [7] gave a lower bound on the success probability.

In their paper [9], Koetter and Médard introduced an algebraic geometric point view on network coding. As explained in [3], computational problems in algebraic geometry can often be solved by Gröbner bases. In this paper, we shall show that the exact computation of the minimum q can be made by applying the division algorithm for multivariate polynomials, and we will show that improved estimates for the success probability can be found by applying the footprint bound from Gröbner basis theory. These results introduce a new approach to network coding study. As the improved estimates take into account which monomials occur in the support of the determinant of a certain matrix [7] we study this matrix in details at the end of the paper.

2   Preliminary

We can determine whether or not all sinks can recover all the transmitted information by the determinant of some matrix [7]. We shall review the definition of such determinant. Let $G = (V, E)$ be an directed acyclic graph with possible parallel edges that represents the network topology. The set of source and sink nodes is denoted by S and T respectively. Assume that the source nodes S together get h symbols in $\mathbf{F}_q$ per unit time and try to send them.

Identify the edges in E with the integers $1, \ldots, |E|$. For an edge $j = (u, v)$ we write $\mathrm{head}(j) = v$ and $\mathrm{tail}(j) = u$. We define the $|E| \times |E|$ matrix $F = (f_{i,j})$ where $f_{i,j}$ is a variable if $\mathrm{head}(i) = \mathrm{tail}(j)$ and $f_{i,j} = 0$ otherwise. The variable $f_{i,j}$ is the coding coefficient from i to j.

Index h symbols in $\mathbf{F}_q$ sent by S by $1, \ldots, h$. We also define an $h \times |E|$ matrix $A = (a_{i,j})$ where $a_{i,j}$ is a variable if the edge j is an outgoing edge from the source $s \in S$ sending the i-th symbol and $a_{i,j} = 0$ otherwise. Variables $a_{i,j}$ represent how the source nodes send information to their outgoing edges.

Let $X(l)$ denote the l-th symbol generated by the sources S, and let $Y(j)$ denote the information sent along edge j. The model is described by the following relation

$$Y(j) = \sum_{i=1}^{h} a_{i,j} X(i) + \sum_{i:\mathrm{head}(i)=\mathrm{tail}(j)} f_{i,j} Y(i) .$$

For each sink $t \in T$ define an $h \times |E|$ matrix B_t whose (i, j) entry $b_{t,i,j}$ is a variable if $\mathrm{head}(j) = t$ and equals 0 otherwise. The index i refers to the i-th symbol sent by one of the sources. Thereby variables $b_{t,i,j}$ represent how the sink t process the received data from its incoming edges.

The sink t records the vector

$$\boldsymbol{b}^{(t)} = \left(b_1^{(t)}, \ldots, b_h^{(t)}\right)$$

where

$$b_i^{(t)} = \sum_{j:\text{head}(j)=t} b_{t,i,j} Y(j) \,.$$

We now recall from [7] under which conditions all informations sent by the sources can always be recovered at all sinks. As in [7] we define the Edmonds matrix M_t for $t \in T$ by

$$M_t = \begin{pmatrix} A & 0 \\ I - F & B_t^T \end{pmatrix} \,. \tag{1}$$

Define the polynomial P by

$$P = \prod_{t \in T} |M_t| \,. \tag{2}$$

P is a multivariate polynomial in variables $f_{i,j}$, $a_{i,j}$ and $b_{t,i,j}$. Assigning a value in $\mathbf{F}_q$ to each variable corresponds to choosing a coding scheme. Plugging the assigned values into P gives an element $k \in \mathbf{F}_q$. The following theorem from [7] tells us when the coding scheme can be used to always recover the information generated at the sources S at all sinks in T.

Theorem 1. *Let the notation and the network coding model be as above. Assume a coding scheme has been chosen by assigning values to the variables $f_{i,j}$, $a_{i,j}$ and $b_{t,i,j}$. Let k be the value found by plugging the assigned values into P. Every sink $t \in T$ can recover from $\boldsymbol{b}^{(t)}$ the informations $X(1), \ldots, X(h)$ no matter what they are, if and only if $k \neq 0$ holds.*

Proof. See [7]. □

3   Computation of the Minimum Field Size

We shall study computation of the minimum symbol size q. For this purpose we will need the division algorithm for multivariate polynomials [3, Sec. 2.3] to produce the remainder of a polynomial $F(X_1, \ldots, X_n)$ modulo $(X_1^q - X_1, \ldots, X_n^q - X_n)$ (this remainder is independent of the choice of monomial ordering). We adapt the standard notation for the above remainder which is

$$F(X_1, \ldots, X_n) \text{ rem } (X_1^q - X_1, \ldots, X_n^q - X_n) \,.$$

The reader unfamiliar with the division algorithm can think of the above remainder of $F(X_1, \ldots, X_n)$ as the polynomial produced by the following procedure. As long as we can find an X_i such that X_i^q divides some term in the polynomial under consideration we replace the factor X_i^q with X_i wherever it occurs. The process continues until the X_i-degree is less than q for all $i = 1, \ldots, n$. It is clear that the above procedure can be efficiently implemented.

Proposition 1. *Let $F(X_1, \ldots, X_n)$ be an n-variate polynomial over $\mathbf{F}_q$. There exists an n-tuple $(x_1, \ldots, x_n) \in \mathbf{F}_q^n$ such that $F(x_1, \ldots, x_n) \neq 0$ if and only if*

$$F(X_1, \ldots, X_n) \text{ rem } (X_1^q - X_1, \ldots, X_n^q - X_n) \neq 0 \,.$$

Proof. As $a^q = a$ for all $a \in \mathbf{F}_q$ it holds that $F(X_1, \ldots, X_n)$ evaluates to the same as $R(X_1, \ldots, X_n) := F(X_1, \ldots, X_n) \text{ rem } (X_1^q - X_1, \ldots, X_n^q - X_n)$ in every $(x_1, \ldots, x_n) \in \mathbf{F}_q^n$. If $R(X_1, \ldots, X_n) = 0$ therefore $F(X_1, \ldots, X_n)$ evaluates to zero for every choice of $(x_1, \ldots, x_n) \in \mathbf{F}_q^n$. If $R(X_1, \ldots, X_n)$ is nonzero we consider it first as a polynomial in $\mathbf{F}_q(X_1, \ldots, X_{n-1})[X_n]$ (that is, a polynomial in one variable over the quotient field $\mathbf{F}_q(X_1, \ldots, X_{n-1})$). But the X_n-degree is at most $q - 1$ and therefore it has at most $q - 1$ zeros. We conclude that there exists an $x_n \in \mathbf{F}_q$ such that $R(X_1, \ldots, X_{n-1}, x_n) \in \mathbf{F}_q[X_1, \ldots, X_{n-1}]$ is nonzero. Continuing this way we find $(x_1, \ldots, x_n)$ such that $R(x_1, \ldots, x_n)$ and therefore also $F(x_1, \ldots, x_n)$ is nonzero.   □

From [7, Th. 2] we know that for all prime powers q greater than $|T|$ linear network coding is possible. It is now straightforward to describe an algorithm that finds the smallest field $\mathbf{F}_q$ of prescribed characteristic p for which linear network coding is feasible. We first reduce the polynomial P from (2) modulo the prime p. We observe that although P is a polynomial in all the variables $a_{i,j}$, $b_{t,i,j}$, $f_{i,j}$ the variable $b_{t,i,j}$ appears at most in powers of 1. This is so as it appears at most in a single entry in M_t and does not appear elsewhere. Therefore $\mathbf{F}_q$ can be used for network coding if $P \text{ rem } p$ does not reduce to zero modulo the polynomials $a_{i,j}^q - a_{i,j}$, $f_{i,j}^q - f_{i,j}$. To decide the smallest field $\mathbf{F}_q$ of characteristic p for which network coding is feasible we try first $\mathbf{F}_q = \mathbf{F}_p$. If this does not work we then try $\mathbf{F}_{p^2}$ and so on. To find an $\mathbf{F}_q$ that works we need at most to try $\lfloor \log_p(|T|) \rfloor$ different fields as we know that linear network coding is possible whenever $q > |T|$.

Note that once a field $\mathbf{F}_q$ is found such that the network connection problem is feasible the last part of the proof of Proposition 1 describes a simple way of deciding coefficients $(x_1, \ldots, x_n) \in \mathbf{F}_q^n$ that can be used for network coding.

From [4, Sec. 7.1.3] we know that it is an NP-hard problem to find the minimum field size for linear network coding. Our findings imply that it is NP-hard to find the polynomial P in (2).

4 Computation of the Success Probability of Random Linear Network Coding

In random linear network coding we from the beginning fix for a collection

$$K \subseteq \{1, \ldots, h\} \times \{1, \ldots, |E|\}$$

the $a_{i,j}$'s with $(i, j) \in K$ and also we fix for a collection

$$J \subseteq \{1, \ldots, |E|\} \times \{1, \ldots, |E|\}$$

the $f_{i,j}$'s with $(i,j) \in J$. This is done in a way such that there exists a solution to the network connection problem with the same values for these fixed coefficients. A priori of course we let $a_{i,j} = 0$ if the edge j is not emerging from the source sending information i, and also a priori we of course let $f_{i,j} = 0$ if j is not an adjacent downstream edge of i. Besides these a priori fixed values there may be good reasons for also fixing other coefficients $a_{i,j}$ and $f_{i,j}$ [7]. If for example there is only one upstream edge i adjacent to j we may assume $f_{i,j} = 1$. All the $a_{i,j}$'s and $f_{i,j}$'s which have not been fixed at this point are then chosen randomly and independently. All coefficients are to be elements in $\mathbf{F}_q$. If a solution to the network connection problem exists with the $a_{i,j}$'s and the $f_{i,j}$'s specified, it is possible to determine values of $b_{t,i,j}$ at the sinks such that a solution to the network connection problem is given. Let μ be the number of variables $a_{i,j}$ and $f_{i,j}$ chosen randomly. Call these variables $X_1, \ldots, X_\mu$. Consider the polynomial P in (2) and let $\widetilde{P}$ be the polynomial made from P by plugging in the fixed values of the $a_{i,j}$'s and the fixed values of the $f_{i,j}$'s (calculations taking place in $\mathbf{F}_q$). Then $\widetilde{P}$ is a polynomial in $X_1, \ldots, X_\mu$. The coefficients of $\widetilde{P}$ are polynomials in the $b_{t,i,j}$'s over $\mathbf{F}_q$. Finally, define

$$\widehat{P} := \widetilde{P} \text{ rem } (X_1^q - X_1, \ldots, X_\mu^q - X_\mu).$$

The success probability of random linear network coding is the probability that the random choice of coefficients will lead to a solution of the network connection problem[1] as in Section 2. That is, the probability is the number

$$\frac{|\{(x_1, \ldots, x_\mu) \in \mathbf{F}_q^\mu | \widetilde{P}(x_1, \ldots, x_\mu) \neq 0\}|}{q^\mu}$$
$$= \frac{|\{(x_1, \ldots, x_\mu) \in \mathbf{F}_q^\mu | \widehat{P}(x_1, \ldots, x_\mu) \neq 0\}|}{q^\mu}. \tag{3}$$

To see the first result observe that for fixed $(x_1, \ldots, x_\mu) \in \mathbf{F}_q^\mu$, $\widetilde{P}(x_1, \ldots, x_\mu)$ can be viewed as a polynomial in the variables $b_{t,i,j}$'s with coefficients in $\mathbf{F}_q$ and recall that the $b_{t,i,j}$'s occur in powers of at most 1. Therefore, if $\widetilde{P}(x_1, \ldots, x_\mu) \neq 0$, then by Proposition 1 it is possible to choose the $b_{t,i,j}$'s such that if we plug them into $\widetilde{P}(x_1, \ldots, x_\mu)$ then we get nonzero. The last result follows from the fact that $\widetilde{P}(x_1, \ldots, x_\mu) = \widehat{P}(x_1, \ldots, x_\mu)$ for all $(x_1, \ldots, x_\mu) \in \mathbf{F}_q^\mu$. In this section we shall present a method to estimate the success probability using Gröbner basis theoretical methods.

We briefly review some basic definitions and results of Gröbner bases. See [3] for a more detailed exposition. Let $\mathcal{M}(X_1, \ldots, X_n)$ be the set of monomials in the variables $X_1, \ldots, X_n$. A monomial ordering $\prec$ is a total ordering on $\mathcal{M}(X_1, \ldots, X_n)$ such that

$$L \prec M \Longrightarrow LN \prec MN$$

---

[1] This corresponds to saying that each sink can recover the data at the maximum rate promised by network coding.

holds for all monomials $L, M, N \in \mathcal{M}(X_1, \ldots, X_n)$ and such that every nonempty subset of $\mathcal{M}(X_1, \ldots, X_n)$ has a unique smallest element with respect to $\prec$. The leading monomial of a polynomial F with respect to $\prec$, denoted by $\text{LM}(F)$, is the largest monomial in the support of F. Given a polynomial ideal I and a monomial ordering the footprint $\Delta_\prec(I)$ is the set of monomials that cannot be found as leading monomials of any polynomial in I. The following proposition explains our interest in the footprint (for a proof of the proposition see [2, Pro. 8.32]).

Proposition 2. *Let $\mathbf{F}$ be a field and consider the polynomials $F_1, \ldots, F_s \in \mathbf{F}[X_1, \ldots, X_n]$. Let $I = \langle F_1, \ldots, F_s \rangle \subseteq \mathbf{F}[X_1, \ldots, X_n]$ be the ideal generated by $F_1, \ldots, F_s$. If $\Delta_\prec(I)$ is finite then the number of common zeros of $F_1, \ldots, F_s$ in the algebraic closure of $\mathbf{F}$ is at most equal to $|\Delta_\prec(I)|$.*

Proposition 2 is known as the footprint bound. It has the following corollary.

Corollary 1. *Let $F \in \mathbf{F}[X_1, \ldots, X_n]$ where $\mathbf{F}$ is a field containing $\mathbf{F}_q$. Fix a monomial ordering and let*

$$X_1^{j_1} \cdots X_n^{j_n} = \text{LM}\big(F \text{ rem } (X_1^q - X_1, \ldots, X_n^q - X_n)\big) \ .$$

The number of zeros of F over $\mathbf{F}_q$ is at most equal to

$$q^n - \prod_{v=1}^n (q - j_v) \ . \tag{4}$$

Proof. We have

$$\Delta_\prec(\langle F, X_1^q - X_1, \ldots, X_n^q - X_n \rangle)$$
$$\subseteq \Delta_\prec(\langle \text{LM}(F \text{ rem } (X_1^q - X_1, \ldots, X_n^q - X_n)), X_1^q, \ldots, X_n^q \rangle)$$

and the size of the latter set equals (4). The result now follows immediately from Proposition 2. $\qquad\square$

Theorem 2. *Let as above $\widetilde{P}$ be found by plugging into P some fixed values for the variables $a_{i,j}$, $(i,j) \in K$, and by plugging into P some fixed values for the variables $f_{i,j}$, $(i,j) \in J$, and by leaving the remaining μ variables flexible. Assume as above that there exists a solution to the network connection problem with the same values for these fixed coefficients. Denote by $X_1, \ldots, X_\mu$ the variables to be chosen by random and define $\widehat{P} := \widetilde{P} \text{ rem } (X_1^q - X_1, \ldots, X_\mu^q - X_\mu)$. (Note that if $q > |T|$ then $\widehat{P} = \widetilde{P}$). Consider $\widehat{P}$ as a polynomial in the variables $X_1, \ldots, X_\mu$ and let $\prec$ be any fixed monomial ordering. Writing $X_1^{j_1} \cdots X_\mu^{j_\mu} = \text{LM}(\widehat{P})$ the success probability is at least*

$$q^{-\mu} \prod_{v=1}^\mu (q - j_v) \ . \tag{5}$$

As a consequence the success probability is in particular at least

$$q^{-\mu} \min \left\{ \prod_{i=1}^\mu (q - s_i) \, \Big| \, X_1^{s_1} \cdots X_\mu^{s_\mu} \text{ is a monomial in the support of } \widehat{P} \right\} \ . \tag{6}$$

Proof. Let $\mathbf{F}$ be the quotient field $\mathbf{F}_q(X_1, \ldots, X_\mu)$. The result in (5) now follows by applying Corollary 1 and (3). As the leading monomial of $\widetilde{P}$ is of course a monomial in the support of $\widetilde{P}$ (6) is smaller or equal to (5). □

Remark 1. The condition in Theorem 2 that there exists a solution to the network connection problem with the coefficients corresponding to K and J being as specified is equivalent to the condition that $\widehat{P} \neq 0$.

We conclude this section by mentioning without a proof that Gröbner basis theory tells us that the true success probability can be calculated as

$$q^{-\mu}\left(q^\mu - |\Delta_\prec(\langle \widetilde{P}, X_1^q - X_1, \ldots, X_\mu^q - X_\mu \rangle)|\right) .$$

This observation is however of little value as it seems very difficult to compute the footprint

$$\Delta_\prec(\langle \widetilde{P}, X_1^q - X_1, \ldots, X_\mu^q - X_\mu \rangle)$$

due to the fact that μ is typically a very high number.

5 The Bound by Ho et al.

In [7] Ho et al. gave a lower bound on the success probability in terms of the number of edges j with associated random coefficients[2] $\{a_{i,j}, f_{l,j}\}$. Letting η be the number of such edges [7, Th. 2] tells us that if $q > |T|$ and if there exists a solution to the network connection problem with the same values for the fixed coefficients, then the success probability is at least

$$p_{\mathrm{Ho}} = \left(\frac{q - |T|}{q}\right)^\eta . \qquad (7)$$

The proof in [7] of (7) relies on two lemmas of which we only state the first one.

Lemma 1. *Let η be defined as above. The determinant polynomial of M_t has maximum degree η in the random variables $\{a_{i,j}, f_{l,j}\}$ and is linear in each of these variables.*

Proof. See [7, Lem. 3]. Alternatively the proof can be derived as a consequence of Theorem 3 in Section 7. □

Recall, that the polynomial P in (2) is the product of the determinants $|M_t|$, $t \in T$. Lemma 1 therefore implies that the polynomial $\widetilde{P}$ has at most total degree equal to $|T|\eta$ and that no variable appears in powers of more than $|T|$. The assumption $q > |T|$ implies $\widehat{P} = \widetilde{P}$ which makes it particular easy to see that the same of course holds for $\widehat{P}$. Combining this observation with the following lemma shows that the numbers in (5) and (6) are both at least as large as the number (7).

---

[2] We state Ho et al.'s bound only in the case of delay-free acyclic networks.

Lemma 2. *Let* $\eta, |T|, q \in \mathbf{N}, |T| < q$ *be some fixed numbers. Let* $\mu, x_1, \dots, x_\mu \in \mathbf{N}_0$ *satisfy*

$$0 \le x_1 \le |T|, \dots, 0 \le x_\mu \le |T|$$

and $x_1 + \cdots + x_\mu \le |T|\eta$. *The minimal value of*

$$\prod_{i=1}^{\mu} \left(\frac{q - x_i}{q} \right)$$

(taken over all possible values of $\mu, x_1, \dots, x_\mu$*) is*

$$\left(\frac{q - |T|}{q} \right)^\eta .$$

Proof. Assume μ and $x_1, \dots, x_\mu$ are chosen such that the expression attains its minimal value. Without loss of generality we may assume that

$$x_1 \ge x_2 \ge \cdots \ge x_\mu$$

holds. Clearly, $x_1 + \cdots + x_\mu = |T|\eta$ must hold. If $x_i < |T|$ and $x_{i+1} > 0$ then

$$(q - x_i)(q - x_{i+1}) > (q - (x_i + 1))(q - (x_{i+1} - 1))$$

which cannot be the case. So $x_1 = \cdots = x_\eta = |T|$. The remaining x_j's if any all equal zero. $\square$

6   Examples

In this section we apply the methods from the previous sections to two concrete networks. We will see that the estimate on the success probability of random linear network coding that was described in Theorem 2 can be considerably better than the estimate described in [7, Th. 2]. Also we will apply the method from Section 3 to determine the smallest field of characteristic two for which network coding can be successful.

As random linear network coding is assumed to take place at the nodes in a decentralized manner, one natural choice is to set $f_{i,j} = 1$ whenever the indegree of the end node of edge i is one and j is the downstream edge adjacent to i. Clearly, if j is not a downstream edge adjacent to i we set $f_{i,j} = 0$. Whenever none of the above is the case we may choose $f_{i,j}$ randomly. Also if there is only one source and the outdegree of the source is equal to the number of symbols to be send we may enumerate the edges from the source by the numbers $1, \dots, h$ and set $a_{i,j} = 1$ if $1 \le i = j \le h$ and set $a_{i,j} = 0$ otherwise. This strategy can be generalized also to deal with the case of more sources. In the following two examples we will choose the variables in the manner just described. The network in the first example is taken from [4, Ex. 3.1] whereas the network in the second example is new.

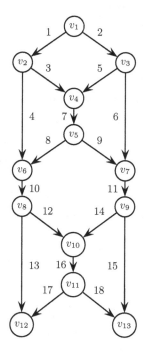

Fig. 1. The network from Example 1

Example 1. Consider the delay-free and acyclic network in Figure 1. There is one sender v_1 and two receivers v_{12} and v_{13}. The min-cut max-flow number is two for both receivers so we assume that two independent random processes emerge from sender v_1. We consider in this example only fields of characteristic 2. Following the description preceding the example we set $a_{1,1} = a_{2,2} = 1$ and $a_{i,j} = 0$ in all other cases. Also we let $f_{i,j} = 1$ except

$$f_{3,7}, f_{5,7}, f_{4,10}, f_{8,10}, f_{9,11}, f_{6,11}, f_{12,16}, f_{14,16}$$

which we choose by random. As in the previous sections we consider $b_{t,i,j}$ as fixed but unknown to us. The determinant polynomial becomes

$$\widetilde{P} = (b^2 c^2 e^2 gh + c^2 f^2 gh + a^2 d^2 f^2 gh)Q \,,$$

where

$$
\begin{array}{llll}
a = f_{3,7} & b = f_{5,7} & c = f_{4,10} & d = f_{8,10} \\
e = f_{9,11} & f = f_{6,11} & g = f_{12,16} & h = f_{14,16}
\end{array}
$$

and $Q = |B'_{v_{12}}||B'_{v_{13}}|$. Here, $B'_{v_{12}}$ respectively $B'_{v_{13}}$ is the matrix consisting of the nonzero columns of $B_{v_{12}}$ respectively the nonzero columns of $B_{v_{14}}$. Restricting to fields $\mathbf{F}_q$ of size at least 4 we have $\widehat{P} = \widetilde{P}$ and we can therefore immediately apply the bounds in Theorem 2. Applying (6) we get the following lower bound on the success probability

$$P_{\text{new 2}}(q) = \frac{(q-2)^3(q-1)^2}{q^5} \, .$$

Choosing as monomial ordering the lexicographic ordering $\prec_{\text{lex}}$ with

$$a \prec_{\text{lex}} b \prec_{\text{lex}} d \prec_{\text{lex}} e \prec_{\text{lex}} g \prec_{\text{lex}} h \prec_{\text{lex}} f \prec_{\text{lex}} c$$

the leading monomial of $\widetilde{P}$ becomes $c^2 f^2 gh$ and therefore from (5) we get the following lower bound on the success probability

$$P_{\text{new 1}}(q) = \frac{(q-2)^2(q-1)^2}{q^4} \, .$$

For comparison the bound (7) from [7] states that the success probability is at least

$$P_{\text{Ho}}(q) = \frac{(q-2)^4}{q^4} \, .$$

We see that $P_{\text{new 1}}$ exceeds P_{Ho} with a factor $(q-1)^2/(q-2)^2$, which is larger than 1. Also $P_{\text{new 2}}$ exceeds P_{Ho}. In Table 1 we list values of $P_{\text{new 1}}(q)$, $P_{\text{new 2}}(q)$ and $P_{\text{Ho}}(q)$ for various choices of q.

Table 1. From Example 1: Estimates on the success probability

q	4	8	16	32	64
$P_{\text{new 1}}(q)$	0.140	0.430	0.672	0.893	0.909
$P_{\text{new 2}}(q)$	0.703×10^{-1}	0.322	0.588	0.773	0.880
$P_{\text{Ho}}(q)$	0.625×10^{-1}	0.316	0.586	0.772	0.880

We next consider the field $\mathbf{F}_2$. We reduce $\widetilde{P}$ modulo $(a^2 - a, \ldots, h^2 - h)$ to get

$$\widehat{P} = (bcegh + cfgh + adfgh)Q \, .$$

From (6) we see that the success probability of random network coding is at least 2^{-5}. Choosing as monomial ordering the lexicographic ordering described above (5) tells us that the success probability is at least 2^{-4}. For comparison the bound (7) does not apply as we do not have $q > |T|$. It should be mentioned that for delay-free acyclic networks the network coding problem is solvable for all choices of $q \geq |T|$ [8] and [11]. From this fact one can only conclude that the success probability is at least 2^{-8} (8 being the number of coefficients to be chosen by random).

Example 2. Consider the network in Figure 2. The sender v_1 generates 3 independent random processes. The vertices v_{11}, v_{12} and v_{13} are the receivers.

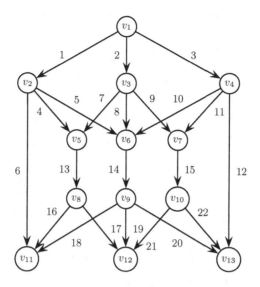

Fig. 2. The network from Example 2

We will apply network coding over various fields of characteristic two. We start by considering random linear network coding over fields of size at least 4. As $4 > |T| = 3$ we know that this can be done successfully.

We set $a_{1,1} = a_{2,2} = a_{3,3} = 1$ and $a_{i,j} = 0$ in all other cases. We let $f_{i,j} = 1$ except $f_{4,13}, f_{7,13}, f_{5,14}, f_{8,14}, f_{10,14}, f_{9,15}, f_{11,15}$, which we choose by random. As in the last section we consider $b_{t,i,j}$ as fixed but unknown to us. Therefore $\widetilde{P} = \widehat{P}$ is a polynomial in the seven variables $f_{4,13}, f_{7,13}, f_{5,14}, f_{8,14}, f_{10,14}, f_{9,15}, f_{11,15}$. The determinant polynomial becomes

$$\widehat{P} = (abcdefg + abce^2f^2 + b^2c^2efg)Q ,$$

where

$$a = f_{4,13} \qquad b = f_{5,14} \qquad c = f_{7,13} \qquad d = f_{8,14}$$
$$e = f_{9,15} \qquad f = f_{10,14} \qquad g = f_{11,15}$$

and $Q = |B'_{v_{11}}||B'_{v_{12}}||B'_{v_{13}}|$. Here, $B'_{v_{11}}$ respectively $B'_{v_{12}}$ respectively $B'_{v_{13}}$ is the matrix consisting of the nonzero columns of $B_{v_{11}}$ respectively the nonzero columns of $B_{v_{12}}$ respectively the nonzero columns of $B_{v_{13}}$. Choosing a lexicographic ordering with d being larger than the other variables and applying (5) we get that the success probability is at least

$$P_{\text{new } 1}(q) = \frac{(q-1)^7}{q^7} .$$

Applying (6) we see that the success probability is at least

$$P_{\text{new } 2}(q) = \frac{(q-1)^3(q-2)^2}{q^5}.$$

For comparison (7) tells us that success probability is at least

$$P_{\text{Ho}}(q) = \frac{(q-3)^3}{q^3} \ .$$

Both bound (5) and bound (6) exceed (7) for all values of $q \geq 4$. In Table 2 we list $P_{\text{new } 1}(q)$, $P_{\text{new } 2}(q)$ and $P_{\text{Ho}}(q)$ for various values of q.

Table 2. From Example 2: Estimates on the success probability

q	4	8	16	32	64
$P_{\text{new } 1}(q)$	0.133	0.392	0.636	0.800	0.895
$P_{\text{new } 2}(q)$	0.105	0.376	0.630	0.799	0.895
$P_{\text{Ho}}(q)$	0.156×10^{-1}	0.244	0.536	0.744	0.865

We next consider the field $\mathbf{F}_2$. We reduce $\widetilde{P}$ modulo $(a^2 - a, \ldots, g^2 - g)$ to get

$$\widehat{P} = (abcdefg + abcef + bcefg)Q \ .$$

From (6) we see that the success probability of random network coding is at least 2^{-7}. Choosing a proper monomial ordering we get from (5) that the success probability is at least 2^{-5}. For comparison neither [7], [8], nor [11] tells us that linear network coding is possible.

7   The Topological Meaning of $|M_t|$

Recall from Section 5 that Ho et al.'s bound (7) relies on the rather rough Lemma 1. The following theorem gives a much more precise description of which monomials can occur in the support of P and $\widetilde{P}$ by explaining exactly which monomials can occur in $|M_t|$. Thereby the theorem gives some insight into when the bounds (5) and (6) are much better than the bound (7). The theorem states that if K is a monomial in the support of $|M_t|$ then it is the product of $a_{i,j}$'s, $f_{i,j}$'s and $b_{t,i,j}$'s related to h edge disjoint paths $P_1, \ldots, P_h$ that originate in the senders and end in receiver t.

Theorem 3. *Consider a delay-free acyclic network. If K is a monomial in the support of the determinant of M_t then it is of the form $K_1 \cdots K_h$ where*

$$K_u = a_{u,l_1^{(u)}} f_{l_1^{(u)},l_2^{(u)}} f_{l_2^{(u)},l_3^{(u)}} \cdots f_{l_{s_u-1}^{(u)},l_{s_u}^{(u)}} b_{t,v_u,l_{s_u}^{(u)}}$$

for $u = 1, \ldots, h$. Here, $\{v_1, \ldots, v_h\} = \{1, \ldots, h\}$ holds and $l_1^{(1)}, \ldots, l_h^{(h)}$ respectively $l_{s_1}^{(1)}, \ldots, l_{s_h}^{(h)}$ are pairwise different. Further

$$f_{l_i^{(u_1)},l_{i+1}^{(u_1)}} \neq f_{l_j^{(u_2)},l_{j+1}^{(u_2)}}$$

unless $u_1 = u_2$ and $i = j$ hold. In other words K corresponds to a product of h edge disjoint paths.

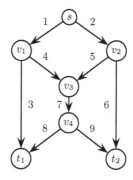

Fig. 3. The butterfly network

Proof. A proof can be found in the appendix. □

We illustrate the theorem with an example.

Example 3. Consider the butterfly network in Figure 3. A monomial K is in the support of $|M_{t_1}|$ if and only if it is in the support of the determinant of

$$N_{t_1} = (n_{i,j}) = \begin{bmatrix} I + F \ B_{t_1}^T \\ A \quad 0 \end{bmatrix}$$

$$= \begin{bmatrix} 1 & 0 & f_{1,3} & f_{1,4} & 0 & 0 & 0 & 0 & 0 & 0 & 0 \\ 0 & 1 & 0 & 0 & f_{2,5} & f_{2,6} & 0 & 0 & 0 & 0 & 0 \\ 0 & 0 & 1 & 0 & 0 & 0 & 0 & 0 & 0 & b_{t_1,1,3} & b_{t_1,2,3} \\ 0 & 0 & 0 & 1 & 0 & 0 & f_{4,7} & 0 & 0 & 0 & 0 \\ 0 & 0 & 0 & 0 & 1 & 0 & f_{5,7} & 0 & 0 & 0 & 0 \\ 0 & 0 & 0 & 0 & 0 & 1 & 0 & 0 & 0 & 0 & 0 \\ 0 & 0 & 0 & 0 & 0 & 0 & 1 & f_{7,8} & f_{7,9} & 0 & 0 \\ 0 & 0 & 0 & 0 & 0 & 0 & 0 & 1 & 0 & b_{t_1,1,8} & b_{t_1,2,8} \\ 0 & 0 & 0 & 0 & 0 & 0 & 0 & 0 & 1 & 0 & 0 \\ a_{1,1} & a_{1,2} & 0 & 0 & 0 & 0 & 0 & 0 & 0 & 0 & 0 \\ a_{2,1} & a_{2,2} & 0 & 0 & 0 & 0 & 0 & 0 & 0 & 0 & 0 \end{bmatrix}$$

By inspection we see that the monomial

$$K = a_{1,1} a_{2,2} b_{t_1,1,3} b_{t_1,2,8} f_{7,8} f_{5,7} f_{2,5} f_{1,3}$$

is in the support of $|N_{t_1}|$. We can write $K = K_1 K_2$ where

$$K_1 = a_{1,1} f_{1,3} b_{t_1,1,3} \quad \text{and} \quad K_2 = a_{2,2} f_{2,5} f_{5,7} f_{7,8} b_{t_1,2,8} \ .$$

This is the description guaranteed by Theorem 3. To make it easier for the reader to follow the proof of Theorem 3 in the appendix we now introduce some of the notations to be used there. By inspection the monomial K can be written

$$K = \prod_{i=1}^{11} n_{i,p(i)}$$

where the permutation p is given by

$$p(1) = 3 \quad p(2) = 5 \quad p(3) = 10 \quad p(4) = 4 \quad p(5) = 7 \quad p(6) = 6$$
$$p(7) = 8 \quad p(8) = 11 \quad p(9) = 9 \quad p(10) = 1 \quad p(11) = 2$$

Therefore if we index the elements in $\{1, \ldots, 11\}$ by

$$i_1 = 10 \quad i_2 = 1 \quad i_3 = 3 \quad i_4 = 11 \quad i_5 = 2 \quad i_6 = 5$$
$$i_7 = 7 \quad i_8 = 8 \quad i_9 = 4 \quad i_{10} = 6 \quad i_{11} = 9$$

then we can write

$$K_1 = n_{i_1, p(i_1)} n_{i_2, p(i_2)} n_{i_3, p(i_3)}$$
$$K_2 = n_{i_4, p(i_4)} n_{i_5, p(i_5)} n_{i_6, p(i_6)} n_{i_7, p(i_7)} n_{i_8, p(i_8)}$$

and we have

$$n_{i_9, p(i_9)} = n_{i_{10}, p(i_{10})} = n_{i_{11}, p(i_{11})} = 1$$

corresponding to the fact $p(i_9) = i_9$, $p(i_{10}) = i_{10}$ and $p(i_{11}) = i_{11}$.

Remark 2. The procedures described in the proof of Theorem 3 can be reversed. This implies that there is a bijective map between the set of edge disjoint paths $P_1, \ldots, P_h$ in Theorem 3 and the set of monomials in $|M_t|$.

Theorem 3 immediately applies to the situation of random network coding if we plug into the $a_{i,j}$'s and into the $f_{t,i,j}$'s on the paths $P_1, \ldots, P_h$ the fixed values wherever such are given. Let as in Lemma 1 η be the number of edges for which some coefficients $a_{i,j}, f_{i,j}$ are to be chosen by random. Considering the determinant as a polynomial in the variables to be chosen by random with coefficients in the field of rational expressions in the $b_{t,i,j}$'s we see that no monomial can contain more than η variables and that no variable occurs more than once. This is because the paths $P_1, \ldots, P_h$ are edge disjoint. Hence, Lemma 1 is a consequence of Theorem 3.

Acknowledgments

The authors would like to thank the anonymous referees for their helpful suggestions.

References

1. Ahlswede, R., Cai, N., Li, S.-Y.R., Yeung, R.W.: Network information flow. IEEE Transactions on Information Theory 46(4), 1204–1206 (2000)
2. Becker, T., Weispfenning, V.: Gröbner Bases - A Computational Approach to Commutative Algebra. Springer, Berlin (1993)
3. Cox, D., Little, J., O'Shea, D.: Ideals, Varieties, and Algorithms, 2nd edn. Springer, Berlin (1996)

4. Fragouli, C., Soljanin, E.: Network Coding Fundamentals. In: Foundations and Trends in Networking, vol. 2(1). now Publishers Inc., Hanover (2007)
5. Geil, O.: On codes from norm-trace curves. Finite Fields and their Applications 9(3), 351–371 (2003)
6. Ho, T., Karger, D.R., Médard, M., Koetter, R.: Network Coding from a Network Flow Perspective. In: Proceedings. IEEE International Symposium on Information Theory, Yokohama, Japan, July 2003, p. 441 (2003)
7. Ho, T., Médard, M., Koetter, R., Karger, D.R., Effros, M., Shi, J., Leong, B.: A Random Linear Network Coding Approach to Multicast. IEEE Transactions on Information Theory 52(10), 4413–4430 (2006)
8. Jaggi, S., Chou, P.A., Jain, K.: Low Complexity Algebraic Multicast Network Codes. In: Proceedings. IEEE International Symposium on Information Theory, Yokohama, Japan, July 2003, p. 368 (2003)
9. Koetter, R., Médard, M.: An Algebraic Approach to Network Coding. IEEE/ACM Transactions on Networking 11(5), 782–795 (2003)
10. Li, S.-Y.R., Yeung, R.W., Cai, N.: Linear Network Coding. IEEE Transactions on Information Theory 49(2), 371–381 (2003)
11. Sanders, P., Egner, S., Tolhuizen, L.: Polynomial Time Algorithms for Network Information Flow. In: Proceedings of the 15th ACM Symposium on Parallel Algorithms, San Diego, USA, June 2003, pp. 286–294 (2003)

A   Proof of Theorem 3

The proof of Theorem 3 calls for the following technical lemma.

Lemma 3. *Consider a delay-free acyclic network with corresponding matrix F as in Section 2. Let I be the $|E| \times |E|$ identity matrix and define*

$$\Gamma = (\gamma_{i,j}) = I + F .$$

Given a permutation p on $\{1, \ldots, |E|\}$ write

$$p^{(i)}(\lambda) = \overbrace{p(p(\cdots(\lambda)\cdots))}^{i \; times}$$

If for some $\lambda \in \{1, \ldots, |E|\}$ the following hold

(1) $\lambda, p(\lambda), \ldots, p^{(x)}(\lambda)$ are pairwise different
(2) $p^{(x+1)}(\lambda) \in \{\lambda, p(\lambda), \ldots, p^{(x)}(\lambda)\}$
(3) $\gamma_{\lambda, p(\lambda)}, \gamma_{p(\lambda), p(p(\lambda))}, \ldots, \gamma_{p^{(x)}(\lambda), p^{(x+1)}(\lambda)}$ are all nonzero

then $x = 0$.

Proof. Let p be a permutation and let x and λ be numbers such that (1), (2) and (3) hold. As p is a permutation then (1) and (2) implies that $p(p^{(x)}(\lambda)) = \lambda$. Aiming for a contradiction assume $x > 0$. As $p(\eta) = \eta$ does not hold for any $\eta \in \{\lambda, p(\lambda), \ldots, p^{(x)}(\lambda)\}$,

$$\gamma_{\lambda, p(\lambda)}, \gamma_{p(\lambda), p^{(2)}(\lambda)}, \ldots, \gamma_{p^{(x)}(\lambda), p^{(x+1)}(\lambda)}$$

are all non-diagonal elements in $I + F$. By (3) we therefore have constructed a cycle in a cycle-free graph and the assumption $x > 0$ cannot be true.   □

Proof (of Theorem 3). A monomial is in the support of the determinant of M_t if and only if it is in the support of the determinant of

$$N_t = \begin{pmatrix} I + F & B_t^T \\ A & 0 \end{pmatrix} = (n_{i,j}) \ .$$

To ease the notation in the present proof we consider the latter matrix. Let p be a permutation on $\{1, \ldots, |E| + h\}$ such that

$$\prod_{s=1}^{|E|+h} n_{s,p(s)} \neq 0 \ . \tag{8}$$

Below we order the elements in $\{1, \ldots, |E| + h\}$ in a particular way by indexing them $i_1, \ldots, i_{|E|+h}$ according to the following set of procedures.

Let $i_1 = |E| + 1$ and define recursively

$$i_s = p(i_{s-1})$$

until $|E| < p(i_s) \leq |E| + h$. Note that this must eventually happen due to Lemma 3. Let s_1 be the (smallest) number such that $|E| < p(i_{s_1}) \leq |E| + h$ holds. This corresponds to saying that $n_{i_1,p(i_1)}$ is an entry in A, that $n_{i_2,p(i_2)}$, $\ldots, n_{i_{s_1-1},p(i_{s_1-1})}$ are entries in $I + F$, and that $n_{i_{s_1},p(i_{s_1})}$ is an entry in B_t^T. Observe, that $p(i_r) = i_r$ cannot happen for $2 \leq r \leq s_1$ as already $p(i_{r-1}) = i_r$ holds. As $n_{i_r,p(i_r)}$ is non-zero by (8) we therefore must have

$$n_{i_r,p(i_r)} = f_{i_r,p(i_r)} = f_{i_r,i_{r+1}}$$

for $2 \leq r < s_1$. Hence,

$$(n_{i_1,p(i_1)}, \ldots, n_{i_{s_1},p(i_{s_1})}) = (a_{1,i_2}, f_{i_2,i_3}, \ldots, f_{i_{s_1-1},i_{s_1}}, b_{t,v_1,i_{s_1}})$$

for some v_1. Denote this sequence by P_1. Clearly, P_1 corresponds to the polynomial K_1 in the theorem.

We next apply the same procedure as above starting with $i_{s_1+1} = |E| + 2$ to get a sequence P_2 of length s_2. Then we do the same with $i_{s_1+s_2+1} = |E| + 3, \ldots, i_{s_1+\cdots s_{h-1}+1} = |E| + h$ to get the sequences $P_3, \ldots, P_h$. For $u = 2, \ldots, h$ we have

$$P_u = \left(n_{i_{s_1+\cdots+s_{u-1}+1},p(i_{s_1+\cdots+s_{u-1}+1})}, \ldots, n_{i_{s_1+\cdots+s_u},p(i_{s_1+\cdots+s_u})} \right)$$

$$= \left(a_{u,i_{s_1+\cdots+s_{u-1}+2}}, f_{i_{s_1+\cdots+s_{u-1}+2},i_{s_1+\cdots+s_{u-1}+3}}, \ldots, \right.$$

$$\left. f_{i_{s_1+\cdots+s_u-1},i_{s_1+\cdots+s_u}}, b_{t,v_u,i_{s_1+\cdots+s_u}} \right) \ .$$

Clearly, P_u corresponds to K_u in the theorem. Note that the sequences $P_1, \ldots, P_h$ by the very definition of a permutation are edge disjoint in the sense that

(1) $n_{i,j}$ occurs at most once in $P_1, \ldots, P_h$,

(2) if n_{j,l_1}, n_{j,l_2} occur in $P_1, \ldots, P_h$ then $l_1 = l_2$,

(3) if $n_{j_1,l}, n_{j_2,l}$ occur in $P_1, \ldots, P_h$ then $j_1 = j_2$.

Having indexed $s_1 + \cdots + s_h$ of the integers in $\{1, \ldots, |E| + h\}$ we consider what is left, namely

$$\Lambda = \{1, \ldots, |E| + h\} \setminus \{i_1, \ldots, i_{s_1 + \ldots + s_h}\} .$$

By construction we have $i_1 = |E| + 1, \ldots, i_{s_1 + \cdots + s_{h-1} + 1} = |E| + h$ and therefore $\Lambda \subseteq \{1, \ldots, |E|\}$. Also by construction for every

$$\delta \in \{1, \ldots, |E|\} \cap \{i_1, \ldots, i_{s_1 + \cdots + s_h}\}$$

we have $\delta = p(\epsilon)$ for some $\epsilon \in \{i_1, \ldots, i_{s_1 + \cdots + s_h}\}$. Therefore $p(\lambda) \in \Lambda$ for all $\lambda \in \Lambda$ holds. In particular $p^{(x)}(\lambda) \in \{1, \ldots, |E|\}$ for all x. From Lemma 3 we conclude that $p(\lambda) = \lambda$ for all $\lambda \in \Lambda$. $\square$

Montgomery Ladder for All Genus 2 Curves in Characteristic 2

Sylvain Duquesne

Université Montpellier II,
Laboratoires I3M, UMR CNRS 5149 and LIRMM, UMR CNRS 5506
Place Eugène Bataillon CC 051, 34005 Montpellier Cedex, France
duquesne@math.univ-montp2.fr

Abstract. Using the Kummer surface, we generalize Montgomery ladder for scalar multiplication to the Jacobian of genus 2 curves in characteristic 2. Previously this method was known for elliptic curves and for genus 2 curves in odd characteristic. We obtain an algorithm that is competitive compared to usual methods of scalar multiplication and that has additional properties such as resistance to simple side-channel attacks. Moreover it provides a significant speed-up of scalar multiplication in many cases. This new algorithm has very important applications in cryptography using hyperelliptic curves and more particularly for people interested in cryptography on embedded systems (such as smart cards).

Keywords: Hyperelliptic curves, Characteristic 2, Kummer surface, Cryptography, Scalar multiplication.

1 Introduction

Elliptic curve cryptosystems were simultaneously introduced by Koblitz [17] and Miller [26]. They are becoming more and more popular because the key length can be chosen smaller than with RSA cryptosystems for the same level of security. This small key size is especially attractive for small cryptographic devices like smart cards. Hyperelliptic curves allow to generalize elliptic curves cryptosystems on smaller base field where basic operations are cheaper. In all schemes, the dominant operation is a scalar multiplication of some point on an elliptic curve (or some element on the Jacobian of a hyperelliptic curve). Hence, the efficiency of this scalar multiplication is central in elliptic and hyperelliptic curve cryptography. In this paper we are dealing with scalar multiplication on Jacobian of genus 2 curves defined over a field of characteristic 2.

For certain elliptic curves, Montgomery [27] developed a method, called Montgomery ladder, allowing faster scalar multiplication than usual methods. His method has the extra advantage that it is resistant to simple side-channel attacks. This is very interesting for people who want to use elliptic curves on embedded devices like smart cards. This method was generalized to any elliptic curves in odd characteristic in [1], to elliptic curves in characteristic 2 [24] and

J. von zur Gathen, J.L. Imaña, and Ç.K. Koç (Eds.): WAIFI 2008, LNCS 5130, pp. 174–188, 2008.

more recently to genus 2 curves in odd characteristic [6,12]. Finally, Gaudry announced last year in [13] he was able to deal with certain genus 2 curves in characteristic 2 and he is currently writing a paper with Lubicz [14].

The aim of this paper is the generalization to all genus 2 curves in characteristic 2. In the following, **K** will denote a field of characteristic 2, M a multiplication in **K**, S a squaring and M_c a multiplication by a constant depending only on the curve. The cost of this operation is usually the same as the one of a multiplication M but it can be neglected if the curve is well chosen. For cryptographic applications, the base field we have in mind is $\mathbb{F}_{2^d}$ where d is a prime number (because of the Weil descent [11]). This paper is organized as follows: in Sections 2 and 3 we recall Montgomery ladder for elliptic curves and basic statements on genus 2 curves. In Section 4 we introduce the Kummer surface in characteristic 2 which allows to develop a Montgomery ladder. Formulas for addition and doubling and comparisons with other formulas are given in Section 5.

2   Montgomery Ladder on Elliptic Curves in Characteristic 2

Let E be an elliptic curve defined over **K** by the equation

$$y^2 + xy = x^3 + a_2 x^2 + a_6,$$

with $a_6 \neq 0$. Every non-supersingular elliptic curve defined over **K** is isomorphic to a curve given by such an equation. The problem we are interested in for cryptographic purposes is the followingscalar multiplication, namely:

Given a point $P \in E(\mathbf{K})$ and an integer n, compute nP as fast as possible.

Of course there are a lot of methods to do this (double and add, sliding window, w-NAF, ...). To improve these algorithms, it is very convenient to use projective coordinates in order to avoid expensive divisions in **K**. Hence, we use a triple (X, Y, Z) in $\mathbb{P}^2(\mathbf{K})$ such that $x = X/Z$ and $y = Y/Z$ to represent the point (x, y).

In [27], Montgomery proposed to avoid computation of the y-coordinate, so that we can hope that basic operations (doubling and addition) are easier to compute. Since, for any x-coordinate, there are two corresponding points on the curve which are opposite, this restriction is equivalent to identifying a point on the curve and its opposite. If we want to add two points $\pm P$ and $\pm Q$, we cannot decide if the result obtained is either $\pm(P + Q)$ as required or $\pm(P - Q)$. So the group law is lost. Nevertheless, some operations remain possible like doubling. Unfortunately, doubling is not sufficient for a complete scalar multiplication. In fact additions are possible if the difference $P - Q$ is known. Then, the principle to compute nP is to use pairs of consecutive multiples of P. The algorithm for scalar multiplication, usually called Montgomery ladder, is as follows:

Algorithm 1. *Montgomery scalar multiplication algorithm on elliptic curves*

Input : $P \in E(\mathbf{K})$ and $n = (n_{\ell-1} \cdots n_0)$ an integer in binary representation.
Output : x and z-coordinate of nP.

1. Initialize $Q = (Q_1, Q_2) = (P, 2P)$
2. for i from $\ell - 2$ down to 1 do
       if $n_i = 0$ then $Q = (2Q_1, Q_1 + Q_2)$
       if $n_i = 1$ then $Q = (Q_1 + Q_2, 2Q_2)$
3. return Q_1

At each step, $Q = (kP, (k+1)P)$ for some k and we compute either $(2kP, (2k+1)P)$ or $((2k+1)P, (2k+2)P)$, so we always have $Q_2 - Q_1 = P$ and additions can be performed.

Contrary to double and add or sliding window methods, both an addition and a doubling are done for each bit of the exponent. It is the price to pay to avoid the y-coordinate but the gain obtained thanks to this restriction is sufficient to compensate for the larger number of operations.

In [24], Lopez and Dahab generalized Montgomery's idea to binary curves and gave formulas for addition (assuming the difference is known) and doubling requiring respectively 4 multiplications and 1 squaring in $\mathbf{K}$ (4M and 1S) and 2 M and 3 S. Thus, the cost of Montgomery ladder is about $6|n|_2$ M and $4|n|_2$ S where $|n|_2$ denotes the number of bits of n. This makes Montgomery ladder one of the best known algorithm for scalar multiplication algorithms in characteristic 2 [32]. Moreover, Montgomery ladder has the extra advantage to be resistant against simple side-channel attacks.

These attacks use observations like timings [18], power consumption [19] or electromagnetic radiation [30]. They are based on the fact that addition and doubling are two different operations on elliptic curves. It is then easy to decide, for each bit of the exponent, if the algorithm (double and add for example) is performing either a doubling (if the bit is 0) or a doubling and an addition (if the bit is 1). Hence, it is easy to recover the whole exponent (which is often the secret key).

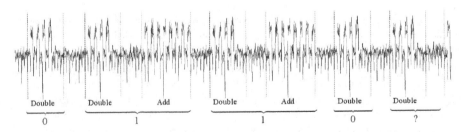

Of course, various countermeasures have been proposed to secure the elliptic curve scalar multiplication against side-channel attacks [5]. For example, if we want to protect a double and add algorithm, the basic way is to perform extra, useless, additions when the bit of the exponent is 0. In this way, for each bit of the exponent we perform both an addition and a doubling so bits of the exponent are indistinguishable. This is of course time consuming and sensible to fault attacks but there are more efficient countermeasures [22].

With Montgomery ladder, we always have to perform both an addition and a doubling for each bit of the exponent, so this method is naturally resistant to simple side-channel attacks. Therefore it is particularly interesting and attractive for people interested in elliptic curve cryptosystems on embedded systems. That is one of the reasons why we want to generalize this method to hyperelliptic curves of genus 2.

Finally, for some cryptosystems (like Diffie-Hellman key exchange, authentication), the x-coordinate of nP is sufficient but others require the y-coordinate. Formulas to recover it are given in [24,29]. However, it is also possible to use variants of these protocols where the x-coordinate is sufficient. Of course these remarks are also valid for hyperelliptic curves [31]. Thereafter we will not take this into consideration.

In order to generalize this method to genus 2 curves, let us first recall some essential background on these curves.

3   Background on Genus 2 Curves in Characteristic 2

Every genus 2 curve is hyperelliptic. So, in the following, we do not state that the curves we are interested in are hyperelliptic.

Moreover, as usual in cryptography, we will concentrate on imaginary curves defined over $\mathbf{K}$ which are given by equations of the form

$$\mathcal{C} : y^2 + h(x)y = f(x), \tag{1}$$

where f and g are in $\mathbf{K}[x]$, $\deg(f) = 5$, $\deg(h) \leq 2$ and there is no singular affine points.

3.1   Arithmetic of Genus 2 Curves

Contrary to elliptic curves, the set of points on genus 2 curves does not form a group. But the Jacobian of $\mathcal{C}$, denoted $\mathcal{J}(\mathcal{C})$, is a group (in the case of elliptic curves, this Jacobian is isomorphic to the curve itself). More details on the definition of the Jacobian can be found in [25] or [5]. There are mainly two ways to represent elements in the Jacobian:

- with a couple of points ($P_1 = (x_1, y_1)$ and $P_2 = (x_2, y_2)$) on the curve such that the unoredered pair (P_1, P_2) is fixed under the Frobenius action (this is a consequence of the Riemann-Roch theorem),
- with 2 polynomials u and v in $\mathbf{K}[x]$ where u is monic, $\deg(v) < \deg(u) \leq 2$ and $u | (v^2 + hv + f)$ (Mumford representation [28]).

The correspondence between these representations is that $u(x) = (x + x_1)(x + x_2)$ and $v(x_i) = y_i$ with appropriate multiplicities. In [3], Cantor described the group law on Jacobians with Mumford representation (in characteristic 0).

Several researchers such as Koblitz [15], or more recently Lange [21] made explicit the steps of Cantor's Algorithm and listed the operations one really needs to perform. They obtained explicit formulas for the group law on the Jacobian which are an analog of the different choices of coordinates for elliptic curves. In this this paper, our purpose is to give an analog of Montgomery ladder. Let us first recall that the equation defining the curve can be simplified.

3.2   Classification of Genus 2 Hyperelliptic Curves over $\mathbb{F}_{2^d}$

As explained in the previous section, we are interested in curves defined over $\mathbb{F}_{2^d}$ by an equation of the form

$$y^2 + (h_2 x^2 + h_1 x + h_0)y = x^5 + f_4 x^4 + f_3 x^3 + f_2 x^2 + f_1 x + f_0. \qquad (2)$$

In fact, genus 2 curves can be divided into three types depending on the leading coefficient of h. Following the notations of [4], we have:

- type I: $h_2 \neq 0$.
- type II: $h_2 = 0$, $h_1 \neq 0$.
- type III: $h_2 = h_1 = 0$, $h_0 \neq 0$.

Moreover, Choie and Yun prove in [4] that type I has asymptotically between $2q^3$ and $4q^3$ isomorphism classes ($q = 2^d$), type II about $2q^2$ and type III between $2q$ and $32q$. However, Galbraith proved in [10] that type III curves are supersingular (hence cryptographically weaker), so only curves of types I and II are interesting for cryptosystems if pairings are not used.

In [2], equations for type I and type II are given in a minimal form, in the sense that if the coefficients range through the base field, the expected number of curves is obtained (say $4q^3$ for type I and $2q^2$ for type II). We recall these minimal forms with slight modifications (we perform a very easy change of variable to assume that $f_1 = 0$ instead of $f_2 = 0$ and we exploit that d is odd). In the following, ε is an element of $\mathbb{F}_2$.

Theorem 1. *A hyperelliptic curve of type I defined over $\mathbb{F}_{2^d}$ with d odd can be rationally transformed into one of the following equations:*

$$type\ Ia: y^2 + (x^2 + h_1 x + h_1^2)y = x^5 + \varepsilon x^4 + f_2 x^2 + f_0,$$
$$type\ Ib: y^2 + (x^2 + h_1 x)y = x^5 + \varepsilon x^4 + f_2 x^2 + f_0.$$

A hyperelliptic curve of type II defined over $\mathbb{F}_{2^d}$ with d odd can be rationally transformed into the following equation :

$$y^2 + xy = x^5 + f_3 x^3 + \varepsilon x^2 + f_0.$$

Remark 1. It is shown in [2] that we can easily find curves of each type suitable for cryptography. Moreover, it will be interesting in the following to choose $h_1 = 1$ for curves of type I. Even with this restriction, there remain sufficiently many isomorphism classes for curves of type I ($4q^2$) and, again, there is no obstruction to find such curves suitable for cryptography.

4 The Kummer Surface in Characteristic 2

With elliptic curves, the main idea of Montgomery method was to avoid the computation of the y-coordinate. It has already been explained in [6] (in odd characteristic) that a good way to generalize this idea is to use the Kummer surface explicitly described in odd characteristic in [9]. This object is the quotient of the Jacobian by the hyperelliptic involution. In other words, we identify an element of the Jacobian and its opposite. The Kummer surface in characteristic $\neq 2$ was known for a long time and we generalized it to fields of characteristic 2 in [7]. It is a quartic surface in $\mathbb{P}^3$. We give here the definition of the Kummer surface and its properties. Proofs and more details can be found in [7]. The Kummer surface is the image of the map

$$\kappa : \mathcal{J}(\mathbf{K}) \longrightarrow \mathbb{P}^3(\mathbf{K})$$
$$\{(x_1, y_1), (x_2, y_2)\} \longmapsto [k_1, k_2, k_3, k_4]$$

with $k_1 = 1, k_2 = x_1 + x_2, k_3 = x_1 x_2$ and

$$k_4 = \frac{(x_1 + x_2)\left(x_1^2 x_2^2 + f_3 x_1 x_2 + f_1\right) + h(x_2) y_1 + h(x_1) y_2}{(x_1 + x_2)^2}$$

Let us note that the image of the point at infinity is $[0, 0, 0, 1]$. More precisely, the Kummer surface is the projective locus given by an equation K of degree four in the first three variables and of degree two in the last one. The exact equation can be found in [7]. In passing from the Jacobian to the Kummer surface, we lost the group structure but traces of it remain. For example, it is possible to double on the Kummer surface. Nevertheless, for general divisors $\mathcal{A}$ and $\mathcal{B}$, we cannot determine values of $k_i(\mathcal{A} + \mathcal{B})$ from values of $k_i(\mathcal{A})$ and $k_i(\mathcal{B})$ since the latter does not distinguish between $\pm\mathcal{A}$ and $\pm\mathcal{B}$, and so not between $\pm(\mathcal{A} + \mathcal{B})$ and $\pm(\mathcal{A} - \mathcal{B})$ (as was already the case with elliptic curves). However values of $k_i(\mathcal{A} + \mathcal{B})k_j(\mathcal{A} - \mathcal{B}) + \varepsilon_{ij} k_i(\mathcal{A} - \mathcal{B})k_j(\mathcal{A} + \mathcal{B})$ are well determined. We have [7].

Theorem 2. *Let $\mathcal{A}$, $\mathcal{B}$ in $\mathcal{J}(\mathbf{K})$ and $\kappa(\mathcal{A})$, $\kappa(\mathcal{B})$ their image in the Kummer surface. Then, for $i, j \in \{1, \ldots, 4\}$, there are explicit polynomials φ_{ij} biquadratic in $k_i(\mathcal{A}), k_i(\mathcal{B})$ such that projectively*

$$k_i(\mathcal{A} + \mathcal{B})k_j(\mathcal{A} - \mathcal{B}) + \varepsilon_{ij} k_i(\mathcal{A} - \mathcal{B})k_j(\mathcal{A} + \mathcal{B}) = \varphi_{ij}(\mathcal{A}, \mathcal{B}) \qquad (3)$$

where $\varepsilon_{ij} = 1$ if $i \neq j$ and 0 if $i = j$.

These traces of the group law allow to give an analog for genus 2 curves of Montgomery ladder. Let $\mathcal{D}$ be an element of the Jacobian. Our purpose is the computation of $n\mathcal{D}$ for some integer n. As it is not possible to add two divisors except if their difference is known (in which case we will abusively denote by $+$ this "differential addition" on the Kummer surface), the principle is (as for elliptic curves) to use pairs of consecutive multiples of $\mathcal{D}$, so that the difference between the two components of the pair is always known and equal to $\mathcal{D}$ (in fact, it is sufficient to know $\kappa(\mathcal{D})$, the difference in the Kummer surface). The algorithm for scalar multiplication is as follows:

Algorithm 2. *Montgomery scalar multiplication algorithm for genus 2 curves*

Input : $\mathcal{D} \in \mathcal{J}(\mathbf{K})$ and $n = (n_{\ell-1} \cdots n_0)$ an integer in binary representation. Output : $\kappa(n\mathcal{D})$, the image in the Kummer surface of $n\mathcal{D}$.

1. Initialize $(\mathcal{A}, \mathcal{B}) = (\kappa(\mathcal{D}), \kappa(2\mathcal{D}))$
2. for i from $\ell - 2$ down to 1 do
      if $n_i = 0$ then $(\mathcal{A}, \mathcal{B}) = (2\mathcal{A}, \mathcal{A} + \mathcal{B})$
      if $n_i = 1$ then $(\mathcal{A}, \mathcal{B}) = (\mathcal{A} + \mathcal{B}, 2\mathcal{B})$
3. return $\mathcal{A}$

Let us now explain more precisely how the addition and the doubling can be derived from the biquadratic forms and give explicit formulas.

5   Formulas for Addition and Doubling

Assuming the knowledge of $(k_i(\mathcal{A} - \mathcal{B}))_{i=1..4}$, it can be easily seen that the biquadratic forms allow to compute $(k_i(\mathcal{A} + \mathcal{B}))_{i=1..4}$. We can also compute $k_i(2\mathcal{A})$ by putting $\mathcal{A} = \mathcal{B}$.

Proposition 1. *Let $\mathbf{K}$ be a field of characteristic 2 and let $\mathcal{C}$ be a curve of genus 2 defined over $\mathbf{K}$ by an equation of the form (2). Let $\mathcal{A}, \mathcal{B} \in \mathcal{J}(\mathcal{C})$ and $\kappa(\mathcal{A}) = [k_1(\mathcal{A}), k_2(\mathcal{A}), k_3(\mathcal{A}), k_4(\mathcal{A})], \kappa(\mathcal{B}) = [k_1(\mathcal{B}), k_2(\mathcal{B}), k_3(\mathcal{B}), k_4(\mathcal{B})]$ their images in the Kummer surface. Assume that the difference $\mathcal{A} - \mathcal{B}$ is known and that $k_1(\mathcal{A} - \mathcal{B}) = 1$ (remember we are in $\mathbb{P}^3(\mathbf{K})$). Then we obtain the Kummer coordinates for $\mathcal{A} + \mathcal{B}$ by the following formulas :*

$$k_1(\mathcal{A} + \mathcal{B}) = \varphi_{11}(\mathcal{A}, \mathcal{B})$$
$$k_2(\mathcal{A} + \mathcal{B}) = \varphi_{12}(\mathcal{A}, \mathcal{B}) + k_1(\mathcal{A} + \mathcal{B}) \times k_2(\mathcal{A} - \mathcal{B})$$
$$k_3(\mathcal{A} + \mathcal{B}) = \varphi_{13}(\mathcal{A}, \mathcal{B}) + k_1(\mathcal{A} + \mathcal{B}) \times k_3(\mathcal{A} - \mathcal{B})$$
$$k_4(\mathcal{A} + \mathcal{B}) = \varphi_{14}(\mathcal{A}, \mathcal{B}) + k_1(\mathcal{A} + \mathcal{B}) \times k_4(\mathcal{A} - \mathcal{B}).$$

The formulas for $2\mathcal{A}$ are

$$k_1(2\mathcal{A}) = \varphi_{14}(\mathcal{A}, \mathcal{A})$$
$$k_2(2\mathcal{A}) = \varphi_{24}(\mathcal{A}, \mathcal{A})$$
$$k_3(2\mathcal{A}) = \varphi_{34}(\mathcal{A}, \mathcal{A})$$
$$k_4(2\mathcal{A}) = \varphi_{44}(\mathcal{A}, \mathcal{A}).$$

The expressions of the φ_{ij} are given in [7] and are available on the web page of the author [8]. However, they require a large number of operations for a curve given by the general equation (2). The main difficulty is to find expressions which require the least possible number of multiplications in $\mathbf{K}$. In order to provide formulas as efficient as possible, we distinguish the types of curve. In each case, we give more precise expressions of the φ_{ij} we are interested in. For clarity we denote $\kappa(\mathcal{A}) = (k_1, k_2, k_3, k_4)$ and $\kappa(\mathcal{B}) = (l_1, l_2, l_3, l_4)$. We also use the sign $\times$ to denote multiplications that must be done and nothing for multiplications already done before in the formulas. We also choose to provide compact formulas but step-by-step (verified) formulas are given on the web page of the author [8].

5.1   Formulas for Curves of Type Ia

Remember that any curve of type Ia can be defined by an equation of the form

$$y^2 + (x^2 + h_1 x + h_1^2)y = x^5 + \varepsilon x^4 + f_2 x^2 + f_0.$$

Addition and doubling can be computed using the following formulas for φ_{ij} (be carefull that these formulas must be combined with proposition 1 to give the complete differential addition):

$\varphi_{11}(\mathcal{A}, \mathcal{B}) = (k_1 \times l_4 + k_2 \times l_3 + k_4 \times l_1 + k_3 \times l_2)^2$

$\varphi_{12}(\mathcal{A}, \mathcal{B}) = \big(k_1 l_4 + k_4 l_1 + h_1 \times k_1 \times l_3 + h_1 \times k_3 \times l_1 + h_1^2 \times k_1 \times l_2 + h_1^2 \times k_2 \times l_1\big) \times$
$\qquad\qquad\quad (k_2 l_3 + k_3 l_2 + h_1 k_1 l_3 + h_1 k_3 l_1 + h_1^2 k_1 k_2 + h_1^2 k_2 l_1)$

$\varphi_{13}(\mathcal{A}, \mathcal{B}) = h_1^2 \times (k_4 l_1 + k_1 l_4) \times (h_1 k_1 \times h_1 l_1 + k_2 \times l_2) + (h_1 k_3 l_1 + k_4 l_1) \times (k_3 \times l_3 +$
$\qquad\qquad\quad h_1 \times (h_1 k_3 l_1 + k_2 l_3 + h_1^2 k_1 l_2)) + (h_1 k_1 l_3 + k_1 l_4) \times (k_3 l_3 + h_1 \times (h_1 k_1 l_3 +$
$\qquad\qquad\quad k_3 l_2 + h_1^2 k_2 l_1)) + (h_1^2 k_2 l_1 + h_1^2 k_1 l_2) \times \alpha$

$\varphi_{14}(\mathcal{A}, \mathcal{B}) = (\alpha + h_1 \times k_1 l_4)^2 + k_3 l_2 \times (h_1^2 \times (k_4 l_1 + h_1^2 k_2 l_1 + h_1^2 k_1 l_2 + k_2 l_3) + \beta) +$
$\qquad\qquad\quad (k_4 l_1 + k_1 l_4 + k_2 l_3) \times (h_1^2 \times (k_1 l_4 + h_1^2 k_2 l_1 + h_1^2 k_1 l_2) + \beta)$

$\quad$ with $\alpha = h_1 \times (k_2 l_3 + k_3 l_2 + h_1 k_1 l_3 + h_1 k_3 l_1) + k_3 l_3 + h_1^2 \times k_2 l_2$
$\qquad\quad \beta = h_1 \times (k_3 l_3 + h_1^2 k_2 l_2)$

$\varphi_{14}(\mathcal{A}, \mathcal{A}) = \alpha + \beta^2$

$\varphi_{24}(\mathcal{A}, \mathcal{A}) = h_1 \times ((h_1^2 \times k_1^2 + k_2^2) \times (d_1 \times k_1^2 + h_1^2 \times d_2 \times k_1^2 + k_4^2) + (d_2 k_1^2 + h_1^2 \times k_2^2) \times \beta +$
$\qquad\qquad\quad \varphi_{14}(\mathcal{A}, \mathcal{A}))$

$\varphi_{34}(\mathcal{A}, \mathcal{A}) = \beta \times (k_4^2 + h_1^2 d_2 k_1^2) + h_1^2 \times (d_1 k_1^2 \times (h_1^2 k_1^2 + \frac{1}{h_1^2} \times k_3^2) + \alpha)$

$\varphi_{44}(\mathcal{A}, \mathcal{A}) = h_1 \times (h_1^2 k_2^2 \times (d_1 k_1^2 + h_1^2 d_2 k_1^2 + c_2 \times k_2^2) + k_3^2 \times (k_4^2 + d_1 k_1^2 + c_3 \times k_3^2)) +$
$\qquad\qquad\quad (c_1 \times k_1^2 + k_4^2)^2$

$\quad$ with $\alpha = h_1^2 \times (k_1^2 \times k_4^2 + k_2^2 \times k_3^2)$
$\qquad\quad \beta = k_3^2 + h_1^2 k_2^2$

and the precomputed constants

$d_1 = f_0 + \varepsilon h_1^4, \quad d_2 = f_2 + \varepsilon h_1^2 + h_1^3, \quad c_1 = \sqrt{d_1^2 + h_1^2(f_0 + h_1^2 f_2)d_2}, \quad c_2 = \frac{d_1 + h_1^2 d_2 + h_1^5}{h_1} \quad$ and $\quad c_3 = \frac{d_2}{h_1}$

Assuming these constants (only depending on the curve) and the inverse of h_1^2 are precomputed, an addition requires 21M, 13M$_c$ and 2S whereas a doubling requires 9M, 12M$_c$ and 6S. This is, of course, not satisfying but this is the worst case. Moreover, we observe that there are many multiplications by h_1 so it is interesting to choose a curve with $h_1 = 1$. In this case an addition can be done in 21M and 2S and a doubling in 9M, 5M$_c$ and 6S which is much better.

5.2 Formulas for Curves of Type Ib

Remember that any curve of type Ib can be defined by an equation of the form

$$y^2 + (x^2 + h_1 x)y = x^5 + \varepsilon x^4 + f_2 x^2 + f_0.$$

Addition and doubling can be computed using the following formulas for φ_{ij} (be carefull that these formulas must be combined with proposition 1 to give the complete differential addition):

$$\varphi_{11}(\mathcal{A}, \mathcal{B}) = (k_1 \times l_4 + k_2 \times l_3 + k_4 \times l_1 + k_3 \times l_2)^2$$
$$\varphi_{12}(\mathcal{A}, \mathcal{B}) = (k_1 l_4 + k_4 l_1 + h_1 \times k_1 \times l_3 + h_1 \times k_3 \times l_1) \times (k_2 l_3 + k_3 l_2 + h_1 k_1 l_3 + h_1 k_3 l_1)$$
$$\varphi_{13}(\mathcal{A}, \mathcal{B}) = (k_4 l_1 + h_1 k_3 l_1) \times \alpha + (k_1 l_4 + h_1 k_1 l_3) \times \beta$$
$$\varphi_{14}(\mathcal{A}, \mathcal{B}) = \alpha \times \beta$$
$$\text{with } \alpha = h_1 \times (k_1 l_4 + k_2 l_3) + k_3 \times l_3$$
$$\beta = h_1 \times (k_4 l_1 + k_3 l_2) + k_3 l_3$$

$$\varphi_{14}(\mathcal{A}, \mathcal{A}) = h_1^2 \times \left((k_1^2 + k_2^2) \times (k_3^2 + k_4^2) + k_1^2 \times k_3^2 + k_2^2 \times k_4^2 \right) + k_3^4$$
$$\varphi_{24}(\mathcal{A}, \mathcal{A}) = \alpha \times k_1^2 + h_1 \times \left(k_2^2 k_4^2 + k_3^4 + d_1 \times k_1^2 k_3^2 \right)$$
$$\varphi_{34}(\mathcal{A}, \mathcal{A}) = h_1 \times \alpha k_1^2 + h_1^2 \times k_3^4 + k_3^2 \times k_4^2,$$
$$\varphi_{44}(\mathcal{A}, \mathcal{A}) = h_1 \times \left(h_1 \alpha k_1^2 + k_3^2 k_4^2 \right) + \left(k_4^2 + c_1 \times k_1^2 + c_2 \times k_2^2 \right)^2 + c_3 \times k_3^4$$
$$\text{with } \alpha = f_0 h_1 \times \left(h_1^2 \times k_1^2 + k_2^2 + \frac{1}{h_1^2} \times k_3^2 \right)$$

with the precomputed constants

$$d_1 = f_2 + \varepsilon h_1^2 + h_1^3 + \frac{f_0}{h_1^2}, c_1 = \sqrt{f_0 h_1^2 d_1}, c_2 = \sqrt{f_0 h_1^2} \text{ and } c_3 = f_2 + \varepsilon h_1^2$$

Assuming these constants and $\frac{1}{h_1^2}$ are precomputed, an addition requires 14M, $4M_c$ and 1S whereas a doubling requires 5M, $12M_c$ and 6S. This is more satisfying than type Ia. We again observe that there are many multiplications by h_1. So, if $h_1 = 1$ an addition can be done in only 14M and 1S and a doubling in 5M, $5M_c$ and 6S. This becomes interesting and competitive with other methods. Indeed, the best known method for classical scalar multiplication algorithms requires 35M and 6S for doubling (33M and 6S if $h_1 = 1$) whereas our method requires 35M and 7S (24M and 7S if $h_1 = 1$) for both a doubling and an addition, so it is inevitably better. In fact, we can do even better with curves of type II.

5.3 Formulas for Curves of Type II

Remember that any curve of type II can be defined by an equation of the form

$$y^2 + xy = x^5 + f_3 x^3 + \varepsilon x^2 + f_0.$$

Addition and doubling can be computed using the following formulas for φ_{ij} (be carefull that these formulas must be combined with proposition 1 to give the complete differential addition):

$$\varphi_{11}(\mathcal{A}, \mathcal{B}) = (\alpha + \beta)^2$$
$$\varphi_{12}(\mathcal{A}, \mathcal{B}) = (k_1 \times l_3 + k_3 \times l_1)^2$$
$$\varphi_{13}(\mathcal{A}, \mathcal{B}) = k_3 l_1 \times \alpha + k_1 l_3 \times \beta$$
$$\varphi_{14}(\mathcal{A}, \mathcal{B}) = \alpha \times \beta$$
$$\text{with } \alpha = k_1 \times l_4 + k_2 \times l_3$$
$$\beta = k_4 \times l_1 + k_3 \times l_2$$

$$\varphi_{14}(\mathcal{A}, \mathcal{A}) = k_1^2 \times k_4^2 + k_2^2 \times k_3^2$$
$$\varphi_{24}(\mathcal{A}, \mathcal{A}) = k_1^2 \times k_3^2$$
$$\varphi_{34}(\mathcal{A}, \mathcal{A}) = \left(k_3^2 + \sqrt{f_0} \times k_1^2\right)^2$$
$$\varphi_{44}(\mathcal{A}, \mathcal{A}) = \left(f_3 \times \left(k_3^2 + \sqrt{f_0} k_1^2\right) + \sqrt{f_0} \times k_2^2 + k_4^2\right)^2$$

These formulas are very simple and easy to evaluate. There is no doubt that they provide very fast arithmetic on genus 2 curves in characteristic 2. Indeed, assuming $\sqrt{f_0}$ is precomputed, an addition requires 12M and 2 S whereas a doubling requires 3M, $3M_c$ and 6 S. This means that a complete scalar multiplication using Montgomery ladder requires only 18M (and even 15 if f_0 and f_3 are well chosen) and 8 S for each bit of the exponent whereas 20M and 8S are required for the doubling alone in "recent" projective coordinates ([20]). It is more delicate to compare this method with elliptic curve without a fully optimized implementation. However, our method seems to be competitive with Montgomery ladder on elliptic curves (18 multiplications in $\mathbb{F}_{2^d}$ against 6 in $\mathbb{F}_{2^{2d}}$). Let us now give more detailed comparisons.

5.4   Comparison with Usual Algorithms for Scalar Multiplication

To date, the best algorithms for scalar multiplication on genus 2 curves in characteristic 2 are obtained by using mixed projective coordinates or variants [21,20]. In this case, Lange needs around 40 multiplications or squaring both for a mixed addition and for a doubling. However, we can use efficient algorithms (like the sliding window method) whereas, in Montgomery ladder, we must perform both an addition and a doubling for each bit of the exponent. Before comparing efficiencies more precisely, let us put advantages of Montgomery ladder forward.

- As was the case for elliptic curves, Montgomery ladder is naturally resistant to simple side-channel attacks, contrary to other algorithms for scalar multiplications. For this reason it is of great interest to people who need to implement hyperelliptic curves protocols on smart cards or other systems which are sensitive to side-channel attacks.

– This algorithm is very easy to implement, there are no precomputations (as in the sliding window method) and an element on the Kummer surface requires only 4 base field elements whereas weighted projective coordinates require 6 or 8 of them so it is also interesting in terms of memory usage. This is an advantage for constrained environments.

In order to compare efficiency of our method with usual methods, we give in table 1, for each type of curve, complexities for our algorithm and for a sliding window method with window size 4 using the best system of coordinates. In practice, sizes 3 or 4 are used but, for objectivity, we choose to make our comparisons with a lower bound for complexities. The system of coordinates used is the so called "recent coordinate" for type II curves [20] and "new coordinate" for type I curves (except type Ia with $h_1 \neq 1$ where projective coordinates are more appropriate) [21]. There are no distinction between types Ia and Ib in [21], so complexities given here come from [5] and our own counting based on [21] and [2]. These complexities are given for one bit of the exponent (on average for the sliding window method but we do not count the precomputations so the complexities given for sliding window method are underestimated). In order to obtain comparisons as objective as possible, we also give an equivalent in number of multiplication assuming that performing a square in $\mathbb{F}_{2^d}$ is either as expensive as 0.3 multiplications (which is usually the case in polynomial basis) or free (which is the case in normal basis). Finally we give the gain obtained using our method. Note that it is possible to obtain better results if the coefficients of the curve are well chosen (so that multiplication by constants M_c are not taken into account).

Our new method is not interesting in term of efficiency for type Ia curves with $h_1 \neq 1$. In fact we obtain in this case a similar result as in odd characteristic [6]:

Table 1. Comparison between sliding window method and Montgomery ladder

Type of the curve	Ia	Ia ($h_1 = 1$)	Ib	Ib ($h_1 = 1$)	II
Lange's formulas					
Double	38M+7S	33M+6S	37M+6S	33M+6S	20M+8S
Addition	38M+4S	35M+5S	37M+4S	35M+4S	42M+7S
Sliding window	45.6M+7.8S	40M+7S	44.4M+6.8S	40M+6.8S	28.4M+9.4S
with S=0.3M	48M	42.1M	46.5M	42M	31.2M
with S=0	45.6M	40M	44.4M	40M	28.4M
Kummer surface					
Double	21M+6S	13M+6S	17M+6S	10M+6S	6M+6S
Addition	34M+2S	21M+2S	18M+S	14M+S	12M+2S
Montgomery ladder	55M+8S	34M+8S	35M+7S	24M+7S	18M+8S
with S=0.3M	57.4M	36.4M	37.1M	26.1M	20.4M
with S=0	55M	34M	35M	24M	18M
gain with S=0.3M	-19.5%	13.5%	20%	38%	35%
gain with S=0	-20.5%	15%	21%	40%	37%

Montgomery ladder is less efficient than usual scalar multiplication algorithm but is still interesting because of the reasons shown at the beginning of this section (mainly simple side-channel attack resistance). Our method becomes very interesting in term of efficiency for curves of type Ib and II since gain between 20 and 40 percent are obtained. Let us note that for these types of curves, Montgomery ladder cannot be slower that any other scalar multiplication algorithm (with the systems of coordinates known to date of course) since for each bit of the exponent it requires less operations than a doubling alone.

Finally, it is also relevant to compare our algorithm with usual ones in affine coordinate (since inversion are not so expensive in even characteristic) and more precisely with side-channel atomicity developed in [22] (since it also provides simple side-channel resistance). Assuming that the cost of an inversion is 10 multiplications, the algorithm in [22] requires 33M and 5S for both a doubling and an addition whatever the type of the curve. This means that the global cost is around 40 mulitplications for each bit of the exponent which is better than our algorithm only for curves of type Ia with random h_1. In the best case (type II), we win a factor 2 compared to Lange-Mishra alorithm.

5.5   Comparison with Elliptic Curves

The main interest of hyperelliptic cryptosystems is that, for the same level of security, the base field is chosen twice as small as the one used for elliptic cryptosystems. Thus base field operations are cheaper in hyperelliptic cryptosystems. More precisely, for cryptographic applications, the base field arithmetic uses either the school-book or the Karatsuba algorithm so the cost of a base field multiplication in elliptic cryptosystems is at least three times the cost in hyperelliptic cryptosystems. We assume in the following that this minimal ratio holds. However, it is quite delicate to do an objective comparison since this ratio is dependent on the device used (architecture, multiprecision algorithms, size of fields,...). For instance, in constraint environments, doubling the size of the base field can be very expensive and we can take more advantage of our new method in such a case.

It is delicate to decide which is the best known method to perform the scalar multiplication for elliptic curves in characteristic 2 since it also depends on the device (cost of the inversion, use of precomputations, constraints) [32]. For obvious reasons, the more appropriate is to compare our method with the Montgomery ladder for elliptic curves even if it is not the best one (but not so far). This algorithm requires 6 M and 4 S which is equivalent, under our optimistic assumptions, to 18 M and 12 S in $\mathbb{F}_{2^d}$. We can see that this complexity is approximately the same than the one obtained for type II curves (18 M and and 8 S). This leads to a 5 percent gain if S=0.3M. It would be of course more appropriate to perform a real life comparison as the one presented by Bernstein at ECC 2006 in odd characteristic following the work of Gaudry [12]. At present, our team is achieving a library in order to be able to perform this type of comparisons [16] but it is a long term work.

5.6   Comparison with Gaudry-Lubicz Formulas

Recently, in [13], Gaudry announced he was able to generalize his previous work ([12]) to characteristic 2 and he is currently writing a paper with Lubicz on the subject [14]. Their approach is different and uses theta functions to obtain formulas for doubling and differential addition requiring only 4M, $3M_c$ and 5S for the doubling and 11M and 4S for the differential addition. However these formulas can only be used for curves of type Ib with $h_1 = 1$. In this case, Gaudry and Lubicz then obtain a Montgomery ladder complexity which is 20 to 25 % faster than ours (depending on the cost of squarings in $\mathbb{F}_{2^d}$) and which is equivalent to our type II complexity. Unfortunately, their method cannot be used for the other cases. Note again that both Gaudry-Lubicz formulas and our formulas can be improved by a good choice of constants.

6   Conclusion and Prospects

Thanks to the Kummer surface, we generalized the Montgomery ladder to genus 2 curves in characteristic 2. Contrary to odd characteristic, the formulas obtained are competitive for all genus 2 curves (which is important if the curve is chosen by somebody else). This is not so surprising since this was already the case for elliptic curves. All these formulas are available on [8] in magma format. For genus 2 curves of type Ia with $h_1 \neq 1$, our method is less efficient than usual scalar multiplication algorithms on genus 2 curves but has the advantage to be resistant to simple side-channel attacks. This was already the case in odd characteristic so this kind of result was expected. It is more surprising that Montgomery ladder for genus 2 curves provides gains of efficiency between 13 and 40 percent for curves of type Ia (with $h_1 = 1$), Ib and II compared to best algorithms known to date (which are not protected against side-channel attack). On the contrary, our method is less efficient than Gaudry-Lubicz one for curves of type Ib with $h_1 = 1$. We also show that, under some reasonable assumptions, Montgomery ladder for type II curves is competitive with elliptic curves.

Finally we provide an algorithm for scalar multiplication which is not only resistant to side-channel attack but is also very efficient. This proves, if needed, that hyperelliptic curve cryptosystems are a good alternative to elliptic ones.

References

1. Brier, E., Joye, M.: Weierstrass Elliptic Curves and Side-Channel Attacks. In: Naccache, D., Paillier, P. (eds.) PKC 2002. LNCS, vol. 2274. Springer, Heidelberg (2002)
2. Byramjee, B., Duquesne, S.: Classification of genus 2 curves over $\mathbb{F}_{2^n}$ and optimization of their arithmetic. Cryptology ePrint Archive 107 (2004)
3. Cantor, D.G.: Computing on the Jacobian of a hyperelliptic curve. Math. Comp. 48, 95–101 (1987)
4. Choie, Y., Yun, D.: Isomorphism classes of hyperelliptic curves of genus 2 over $\mathbb{F}_q$. In: Batten, L.M., Seberry, J. (eds.) ACISP 2002. LNCS, vol. 2384, pp. 190–202. Springer, Heidelberg (2002)

5. Cohen, H., Frey, G.: Handbook of elliptic and hyperelliptic curve cryptography, Discrete Math. Appl. Chapman & Hall/CRC, Boca Raton (2006)
6. Duquesne, S.: Montgomery scalar multiplication for genus 2 curves. In: Buell, D.A. (ed.) ANTS 2004. LNCS, vol. 3076, pp. 153–168. Springer, Heidelberg (2004)
7. Duquesne, S.: Traces of the group law on the Kummer surface of a curve of genus 2 in characteristic 2, preprint, available at [8]
8. Duquesne, S.: Formulas for traces of the group law on the Kummer surface of a curve of genus 2 in characteristic 2,
   http://www.math.univ-montp2.fr/~duquesne/articles/kummer2
9. Flynn, E.V.: The group law on the Jacobian of a curve of genus 2. J. reine angew. Math. 439, 45–69 (1993)
10. Galbraith, S.: Supersingular curves in cryptography. In: Boyd, C. (ed.) ASIACRYPT 2001. LNCS, vol. 2248, pp. 495–513. Springer, Heidelberg (2001)
11. Gaudry, P., Hess, F., Smart, N.: Constructive and destructive facets of Weil descent on elliptic curves. J. Cryptology 15(1), 19–46 (2002)
12. Gaudry, P.: Fast genus 2 arithmetic based on Theta functions. Journal of Mathematical Cryptology 1, 243–265 (2007)
13. Gaudry, P.: Variants of the Montgomery form based on Theta functions, Toronto (November 2006)
14. Gaudry, P., Lubicz, D.: The arithmetic of characteristic 2 Kummer surfaces. Cryptology ePrint Archive 133 (2008)
15. Harley, R.: Fast arithmetic on genus 2 curves (2000),
   http://cristal.inria.fr/~harley/hyper
16. Imbert, L., Peirera, A., Tisserand, A.: A Library for Prototyping the Computer Arithmetic Level in Elliptic Curve Cryptography. In: Proc. SPIE, vol. 6697, 66970N (2007)
17. Koblitz, N.: Elliptic curve cryptosystems. Math. Comp. 48, 203–209 (1987)
18. Kocher, P.C.: Timing attacks on implementations of DH, RSA, DSS and other systems. In: Koblitz, N. (ed.) CRYPTO 1996. LNCS, vol. 1109, pp. 104–113. Springer, Heidelberg (1996)
19. Kocher, P.C., Jaffe, J., Jun, B.: Differential power analysis. In: Wiener, M.J. (ed.) CRYPTO 1999. LNCS, vol. 1666, pp. 388–397. Springer, Heidelberg (1999)
20. Lange, T.: Arithmetic on binary genus 2 curves suitable for small devices. In: Proceedings ECRYPT Workshop on RFID and Lightweight Crypto., Graz, Austria, July 14-15 (2005)
21. Lange, T.: Formulae for arithmetic on genus 2 hyperelliptic curves. Appl. Algebra Engrg. Comm. Comput. 15(5), 295–328 (2005)
22. Lange, T., Mishra, P.K.: SCA resistant parallel explicit formula for addition and doubling of divisors in the Jacobian of hyperelliptic curves of genus 2. In: Maitra, S., Veni Madhavan, C.E., Venkatesan, R. (eds.) INDOCRYPT 2005. LNCS, vol. 3797, pp. 403–416. Springer, Heidelberg (2005)
23. Lopez, J., Dahab, R.: Improved algorithms for elliptic curve arithmetic in GF(2^n). In: Tavares, S., Meijer, H. (eds.) SAC 1998. LNCS, vol. 1556, pp. 201–212. Springer, Heidelberg (1999)
24. Lopez, J., Dahab, R.: Fast multiplication on elliptic curves over GF(2^m) without precomputation. In: Koç, Ç.K., Paar, C. (eds.) CHES 1999. LNCS, vol. 1717, pp. 316–327. Springer, Heidelberg (1999)
25. Menezes, A., Wu, Y.H., Zuccherato, R.: An elementary introduction to hyperelliptic curves. In: Koblitz, N. (ed.) Algebraic aspects of cryptography. Algorithms and Computation in Mathematics, vol. 3, pp. 155–178 (1998)

26. Miller, V.S.: Use of elliptic curves in cryptography. In: Williams, H.C. (ed.) CRYPTO 1985. LNCS, vol. 218, pp. 417–426. Springer, Heidelberg (1986)
27. Montgomery, P.L.: Speeding the Pollard and elliptic curve methods of factorization. Math. Comp. 48, 164–243 (1987)
28. Mumford, D.: Tata lectures on Theta II. Birkhäuser, Basel (1984)
29. Okeya, O., Sakurai, K.: Efficient Elliptic Curve Cryptosystems from a Scalar Multiplication Algorithm with Recovery of the y-Coordinate on a Montgomery-Form Elliptic Curve. In: Koç, Ç.K., Naccache, D., Paar, C. (eds.) CHES 2001. LNCS, vol. 2162, pp. 126–141. Springer, Heidelberg (2001)
30. Quisquater, J.J., Samyde, D.: ElectroMagnetic Analysis (EMA): Measures and Countermeasures for Smart Cards. In: Attali, S., Jensen, T. (eds.) E-smart 2001. LNCS, vol. 2140, pp. 200–210. Springer, Heidelberg (2001)
31. Smart, N., Siksek, S.: A fast Diffe-Hellman protocol in genus 2. Journal of Cryptology 12, 67–73 (1999)
32. Stam, M.: On Montgomery-Like Representations for Elliptic Curves over $GF(2^k)$. In: Desmedt, Y.G. (ed.) PKC 2003. LNCS, vol. 2567, pp. 240–253. Springer, Heidelberg (2002)

On Cryptographically Significant Mappings over $\mathrm{GF}(2^n)$

Enes Pasalic

IMFM Ljubljana & University of Primorska, Koper
Slovenia
enespasalic@yahoo.se

Abstract. In this paper we investigate the algebraic properties of important cryptographic primitives called substitution boxes (S-boxes). An S-box is a mapping that takes n binary inputs whose image is a binary m-tuple; therefore it is represented as $F : \mathrm{GF}(2)^n \to \mathrm{GF}(2)^m$. One of the most important cryptographic applications is the case $n = m$, thus the S-box may be viewed as a function over $\mathrm{GF}(2^n)$. We show that certain classes of functions over $\mathrm{GF}(2^n)$ do not possess a cryptographic property known as APN (Almost Perfect Nonlinear) permutations. On the other hand, when n is odd, an infinite class of APN permutations may be derived in a recursive manner, that is starting with a specific APN permutation on $\mathrm{GF}(2^k)$, k odd, APN permutations are derived over $\mathrm{GF}(2^{k+2i})$ for any $i \geq 1$. Some theoretical results related to permutation polynomials and algebraic properties of the functions in the ring $\mathrm{GF}(q)[x,y]$ are also presented. For sparse polynomials over the field $\mathrm{GF}(2^n)$, an efficient algorithm for finding low degree I/O equations is proposed.

1 Introduction

Differential cryptanalysis introduced in [2], together with linear cryptanalysis [21] are considered as the most efficient cryptanalyst tools for block ciphers. Commonly, the security of modern block ciphers substantially relies on the cryptographic properties of its substitution boxes (S-boxes), which are in most of the cases the only source of nonlinearity. These S-boxes are most often constructed by means of certain well-known power mappings that have relatively good cryptographic properties such as high nonlinearity, high algebraic degree and good differential characteristics.

However, almost all families of so-called APN (Almost Perfect Nonlinear) functions have been derived from power polynomials, that is $F(x) = x^d$ over the field $\mathrm{GF}(2^n)$ for a suitably chosen d. Following the lines of algebraic attacks, a thorough examination of the known APN power polynomials has been discussed in [8]. It was shown that all classes of known APN power functions are susceptible to a degree decrease by applying some simple transformation techniques. More precisely, the authors consider power polynomials of the form $y = x^d$ for those d which ensure that x^d is APN. Then one applies certain operations such as composition, powering or multiplication to this function in order to obtain

J. von zur Gathen, J.L. Imaña, and Ç.K. Koç (Eds.): WAIFI 2008, LNCS 5130, pp. 189–204, 2008.
© Springer-Verlag Berlin Heidelberg 2008

independent multivariate quadratic equations over $GF(2)$ in x, y. Applying such transformations to the known APN power polynomials it was proved that all classes admit certain number of independent quadratic equations over $GF(2)$ ranging from n (for Niho's and Dobbertin's exponent) to $9n$ for Welch exponent. An exact evaluation on the number of linearly independent bi-affine and quadratic equations (based mainly on the computer simulations for relatively small input size) for the main classes of APN power monomials is given in [10].

Nevertheless, provided with such a number of linearly independent quadratic equations that relate the input to the output of an S-box we still do not have efficient tools for attacking block ciphers based on such S-boxes. For instance the Advanced Encryption Standard (AES) can be represented as a system of quadratic equations over $GF(2)$ [11] or alternatively as an extremely sparse system of quadratic equations over $GF(2^8)$ [22], though there is no technique for solving such a system efficiently. Though solving a random system of quadratic equations (also known as MQ problem [1]) is known to be an NP-hard problem [17], in recent few years a lot of effort has been put to devise efficient algorithms for solving systems of quadratic equations which possess a certain structure [9,16,24]. The structure exploited in these algorithms is commonly the sparseness of the system.

To satisfy the diverse cryptographic criteria, a cryptographically strong S-box can be taken from the class of APN permutations. In addition it should have good algebraic properties, so that low degree I/O (input/output) relations do not exist. Unfortunately, when n is even, it has been a long-term open problem to prove the nonexistence of APN permutations over $GF(2^n)$. We show that certain APN functions, derived from power monomials, cannot be permutations; implying the exclusion of this class of function as a possible candidate in construction of APN permutations.

Ensuring that there do not exist quadratic equations which relate the input and output of a given S-box, the complete cipher might be more resistant to algebraic attacks as the number of monomials in resulting nonquadratic equations is much higher making the complexity of these attacks computationally more demanding. Hence the main question that we try to answer is how do we choose permutation polynomials over finite fields that are not susceptible either to linear or differential cryptanalysis and at the same time have a relatively large algebraic degree which cannot be significantly decreased. In this direction we demonstrate the possibility of constructing APN permutations recursively. Though we cannot provide a rigorous mathematical treatment of the particular recursive class, these functions are very interesting objects for cryptographic applications.

In the last part of this manuscript we propose an efficient algorithm for finding low degree input/output (I/O) relations for sparse polynomials over finite fields. A similar approach has been taken in [8] where several techniques were developed

---

[1] Not all instances of MQ problem are considered hard. That is, solving a set of of m quadratic equations in n variables for a degenerate case $m \ll n$ or $n \ll m$ turns out not to be hard.

for deriving quadratic I/O relations from certain classes of power monomials. Our algorithm takes as input arbitrary sparse polynomial over $\mathbb{F}_{2^n}$ and outputs low degree I/O equations in case of their existence.

The rest of the paper is organized as follows. Section 2 introduces basic definitions and concepts. The nonexistence of certain classes of permutations and the implication of this result to the APN conjecture is treated in Section 3. In this section a recursive construction of APN permutations with good algebraic properties is also discussed. In Section 4 an efficient algorithm for finding low degree I/O relations is given. Section 5 concludes the paper.

2 Preliminaries

In the sequel $\mathbb{F}_{2^n}$ will denote the Galois field of 2^n elements. The *polynomial degree*, denoted by deg_p, associated to $P(x) = \sum_i a_i x^i$ is defined as the largest i for which a_i is nonzero. Any mapping $F : \mathbb{F}_{2^n} \to \mathbb{F}_{2^n}$ can be viewed as a mapping $F' : \mathbb{F}_2^n \to \mathbb{F}_2^n$ by fixing the isomorphism between the vector space $\mathbb{F}_2^n$ and the field $\mathbb{F}_{2^n}$. If we represent the function F as a function on the vector space $\mathbb{F}_2^n$, then we may consider this function as being a collection of n Boolean functions $f_1, \ldots, f_n$, that is, $F' = (f_1, \ldots, f_n)$. Here, the Boolean functions $f_i : \mathbb{F}_2^n \to \mathbb{F}_2$. Then the algebraic degree of F' is defined to be,

Definition 1. *The algebraic degree of F' is defined as,*

$$deg(F') = \min_{\tau \in \mathbb{F}_2^{n*}} deg(\sum_{j=1}^{n} \tau_j f_j(x)), \tag{1}$$

where $deg(f)$ denotes the usual algebraic degree of a Boolean function f, that is, the highest length of the terms that appear in the algebraic normal form of f.

The algebraic degree may also be deduced from the polynomial representation. That is, for a function $F : \mathbb{F}_{2^n} \to \mathbb{F}_{2^n}$ represented as $F(x) = \sum_{i=0}^{2^n-1} a_i x^i$, the algebraic degree is given by

$$deg(F) = \max_i \{wt(i); a_i \neq 0\}, \tag{2}$$

where $wt(i)$ denotes the Hamming weight (number of ones) in a binary representation of integer i. Also for a function $F(x,y) : \mathbb{F}_{2^n} \times \mathbb{F}_{2^n} \to \mathbb{F}_{2^n}$, where $F(x,y) = \sum_{i,j=0}^{2^n-1} a_{i,j} x^i y^j$, the algebraic degree is defined as,

$$deg(F) = \max_{i,j} \{wt(i) + wt(j); a_{i,j} \neq 0\}, \tag{3}$$

The differential properties of $F : \mathbb{F}_{2^n} \to \mathbb{F}_{2^n}$ are visualized through so-called differential table that for each $a \in \mathbb{F}_{2^n}^*, b \in \mathbb{F}_{2^n}$ consists of the number of solutions to the following equation,

$$F(x+a) + F(x) = b \quad a \in \mathbb{F}_{2^n}^*, b \in \mathbb{F}_{2^n}. \tag{4}$$

Then, a function F is called *almost perfect nonlinear* (APN) if each equation (4) has at most two solutions in $\mathbb{F}_{2^n}$ and such a function has a highest resistance to differential cryptanalysis. The differential properties of F are then comprised through the differential table,

$$\{\delta_F(a,b)\} = \{|\{x \in \mathbb{F}_{2^n} : F(x+a) + F(x) = b\}|; a \in \mathbb{F}_{2^n}^*, b \in \mathbb{F}_{2^n}\}.$$

The nonlinearity of $F : \mathbb{F}_{2^n} \to \mathbb{F}_{2^n}$ and hereby the resistance to linear cryptanalysis of Matsui [21] is measured through extended Walsh transform defined as,

$$W_F(\lambda, \gamma) = \sum_{x \in \mathbb{F}_{2^n}} (-1)^{Tr(\gamma F(x) + \lambda x)}, \quad \lambda \in \mathbb{F}_{2^n}, \gamma \in \mathbb{F}_{2^n}^*, \tag{5}$$

where $'Tr'$ denotes the trace mapping, i.e. $Tr(x) = x + x^{2^1} + \cdots + x^{2^{n-1}}$. Then, defining the linearity as

$$\mathcal{L}(F) = \max\{|W_F(\lambda, \gamma)| : \lambda \in \mathbb{F}_{2^n}, \gamma \in \mathbb{F}_{2^n}^*\},$$

the goal is to find mappings with minimum possible value for $\mathcal{L}(F)$. Those F, that achieve the minimum possible value for $\mathcal{L}(F)$ are called AB (*almost bent*) or *maximally nonlinear*, and these functions have the maximum resistance against linear cryptanalysis. For odd $n = 2m + 1$ this value is known to be 2^{m+1} [7]. For even n it is still open problem to determine the minimum for $\mathcal{L}(F)$. The best known value is $2^{n/2+1}$ which is easily obtained from certain power functions x^d.

3 Nonexistence of Certain Classes of Permutations

There are several classes of APN functions for odd n that may be permutations or not, while for even n there exist APN functions but none of these functions is a permutation. It has been highly conjectured that there do not exist APN permutation for even n [15]. The conjecture has been confirmed true for $n = 4$, and for a large class of permutation polynomials $P(x) = \sum_{i=0}^{2^n-1} a_i x^i$, whose coefficients $a_i \in \mathbb{F}_{2^{n/2}}$ [18]. That is, given a permutation polynomial $P(x)$ over $\mathbb{F}_{2^n}$ with coefficents $a_i \in \mathbb{F}_{2^{n/2}}$, it was shown that $P(x)$ is affine on some 2-dimensional subspace, and therefore it cannot be APN [18].

Let x^d be a nonpermutation monomial over $\mathbb{F}_{2^n}$, that is, $\gcd(d, 2^n - 1) = s > 1$. In general, it is an open problem when an arbitrary (non)permuting function G becomes (remains) a permutation when a linear polynomial is added to G. For a special case $F(x) = x^d + L(x)$, where $L(x) = \sum_{k=0}^{n-1} a_k x^{2^k}$ ($a_k \in \mathbb{F}_2$) is a linearized polynomial with binary coefficients, we show that F cannot be a permutation.

Note that taking a nonpermuting binomial $G(x) = x^d + x^s$ where d, s are not 2-power $(d, s \neq 2^i$ for some $i \geq 0)$, we can find permutation polynomials of the form $P(x) = G(x) + L(x)$. One example can be found in [14], where it was proved that for $n = 2m + 1$ the polynomial $P(x) = x^{2^{m+1}+1} + x^3 + x$ is a permutation polynomial on $\mathbb{F}_{2^n}$. The following result shows the nonexistence of certain class of permutation, and therefore it excludes this class as a potential candidate for generation of APN permutations.

Theorem 1. *Let $F(x) = x^d$ over the field $\mathbb{F}_{2^n}$ such that $\gcd(d, 2^n - 1) = s > 1$. Then the polynomial*

$$F(x) + L(x) = F(x) + \sum_{k=0}^{n-1} a_k x^{2^k}, \quad a_k \in \mathbb{F}_2,$$

is never a permutation.

Proof. Let α denote a primitive element in $\mathbb{F}_{2^n}$. It is easy to verify that the mapping $F(x) = x^d$, $\gcd(d, 2^n - 1) = s > 1$, is s-to-one and the image of such a mapping consists of exactly $(2^n - 1)/s$ nonzero values, having the zero point mapped to zero. More precisely we have,

$$S_i = \{\alpha^i, \alpha^{(2^n-1)/s+i}, \alpha^{2(2^n-1)/s+i}, \ldots, \alpha^{(s-1)(2^n-1)/s+i}\} \xrightarrow{x^d} \alpha^{di};$$

for $i = 0, 1, \ldots, (2^n - 1)/s - 1$.

Note that if $F(x) + L(x)$ is to be a permutation then $L(x)$ cannot be a permutation, that is $L(1) = 0$ as otherwise $0, 1 \xrightarrow{F+L} 0$. Then a nonpermuting linearized polynomial L with binary coefficients maps u-to-one, where $u = 2v$, $v \geq 1$. This is because,

$$L(1 + \alpha^i) = L(1) + L(\alpha^i) = L(\alpha^i), \tag{6}$$

which for any α^i gives that $\alpha^i, 1 + \alpha^i \xrightarrow{L} \beta$ for some $\beta \in \mathbb{F}_{2^n}$. The elements $1 + \alpha^i$ and α^i are clearly different for any $0 \leq i \leq 2^n - 2$. There might be some other elements, say α^j, such that $L(\alpha^j) = L(\alpha^i)$, for $i \neq j$. Then we have $L(1 + \alpha^i) = L(\alpha^i) = L(1 + \alpha^j) = L(\alpha^j)$.

Now it is enough to show that the differential equation,

$$(x + 1)^d + x^d = 0, \tag{7}$$

always has at least two solutions. Note that if α^i is a solution of (7) then $1 + \alpha^i$ is a solution as well. Then $F + L$ cannot be a permutation since $F(\alpha^i) + L(\alpha^i) = F(1 + \alpha^i) + L(1 + \alpha^i)$, that is α^i and $\alpha^i + 1$ have the same image.

Since $\gcd(2^n - 1, d) = s > 1$ we may write $d = su$ for some $u \geq 1$ and $\gcd(2^n - 1, u) = 1$. Then the differential equation above may be written as,

$$[(x + 1)^s]^u + [x^s]^u = 0,$$

which is equivalent in terms of solutions to,

$$(x + 1)^s + x^s = 0,$$

as x^u is a permutation on $\mathbb{F}_{2^n}$. This equation is of degree $s - 1$ and it has at most $s - 1$ solutions. Note again that solutions come in pairs, that is, if α^i is a solution so is $1 + \alpha^i$. We show that $(1 + \alpha^{j(2^n-1)/s})^{-1}$ is the set of solutions for $j = 1, \ldots, s - 1$. We have,

$$(x + 1)^s + x^s = 0 \Leftrightarrow \left(\frac{x + 1}{x}\right)^s + 1 = 0 \Leftrightarrow 1 + \frac{1}{x} = \alpha^{j(2^n-1)/s}; \quad j = 1, \ldots, s - 1.$$

This completes the proof.   □

This also implies the nonexistence of APN permutations of the form $x^d + L(x)$, where x^d is a nonpermutation polynomial and L a linearized polynomial with binary coefficients. Note that there is no restriction on the evenness of n.

Corollary 1. *There do not exist APN permutations on $\mathbb{F}_{2^n}$ of the form,*

$$x^d + L(x),$$

where $\gcd(2^n - 1, d) > 1$, *and* $L(x) = \sum_{k=0}^{n-1} a_k x^{2^k}$, $a_k \in \mathbb{F}_2$.

Example 1. *For odd n we know that x^3 is an APN permutation. The APN property of the function x^3 is preserved for even n but then this function is not a permutation, as $3|2^n - 1$. Since x^3 is APN then so is $x^3 + L(x)$ for a linear polynomial with binary coefficients. However the above Theorem guarantees that it cannot be a permutation.*

The general case concerning $L(x)$ with the coefficients from $\mathbb{F}_{2^n}$ seems to be harder to analyze. The difficulty comes from the fact that for $L(x) = \sum_{k=0}^{n-1} a_k x^{2^k}$, $a_k \in \mathbb{F}_{2^n}$, the right-hand side in (6) is not valid as $L(1) = \beta \neq 0$. Then using the same approach as taken in the proof of Theorem 1 would lead to showing the existence of solutions to,

$$(x + \alpha)^d + x^d = \beta,$$

where β is dependent on the choice of the coefficients a_k in $L(x)$. A thorough treatment for the general case of $a_k \in \mathbb{F}_2^k$ is left to the extended version of this paper.

Notice that proving the general case would formally disprove the conjecture stated in [6]. This conjecture claims that given any AB function F there exist a linear function L such that $F + L$ is a permutation. A counterexample for this conjecture has already been found [5], for a certain AB function over $\mathbb{F}_{2^5}$.

Remark 1. The property of being permutation is invariant under the multiplication by a nonzero constant, that is F is a permutation if and only if γF is, $\gamma \in \mathbb{F}_{2^n}^*$. Since the coefficient of $x^d + L(x)$ are binary, the result of Corollary 1 is only a special case of the Hou's result [18] up to the multiplication by a constant.

3.1   An Example of Recursive Construction of APN Permutations

For an arbitrary mapping $F : \mathbb{F}_{2^n} \rightarrow \mathbb{F}_{2^n}$ there always exist I/O equations of degree t (and algorithms for finding these equations), where the value of t is computed as given below [19].

Proposition 1. *For any mapping $y = F(x)$ where $F : \mathbb{F}_2^n \mapsto \mathbb{F}_2^n$ there exists algebraic equation(s) of degree t over $\mathbb{F}_2$ (in indeterminates $x_1, \ldots, x_n, y_1, \ldots, y_n$), where $t < \lceil n/2 \rceil$ is the least positive integer satisfying,*

$$\sum_{i=0}^{t} \binom{2n}{i} > 2^n. \tag{8}$$

Furthermore, a tight lower bound on t is given by, $t \geq \lceil n/4 \rceil + 1$.

For instance, this bound implies that for a symmetric S-box of size $n = 7$ or $n = 8$ there always exist cubic I/O equations which describe given S-box. The inverse S-box of the AES ($n = 8$) admits quadratic I/O equations and therefore it is not optimized with respect to so-called algebraic immunity.

To preserve good differential and linear properties, we may attempt to find instances (or classes) of polynomials derived from power monomials that have better algebraic properties than power monomials. It is well-known that the differential and linear properties of the power permutation $F(x) = x^d$ are the same as for $x^{2d}, x^{4d}, \ldots, x^{2^{n-1}}$ or for the inverse coset $x^{-d}, x^{-2d}, \ldots, x^{-2^{n-1}}$. While $x^{2^i d}$ is of the same algebraic degree as x^d, this is not the case for the inverse cyclotomic coset as $x^{-2^i d}$ has in general different algebraic degree than x^d. The linear and inverse transformation may be then unified in so-called *extended affine equivalence* (EA equivalence) so that F and F' are EA-equivalent if $F' = A_1 \circ F \circ A_2 + A$ for some affine permutations A_1, A_2 and affine function A. The equivalence also includes the inverse coset by replacing F with F^{-1}.

A more general framework was first introduced in [6], where the transformation is performed rather to the graph of functions. Then $F, F' : \mathbb{F}_{2^n} \to \mathbb{F}_{2^n}$ are called CCZ-equivalent, terminology introduced in [5], if the sets $G_F = \{(x, F(x)) | x \in \mathbb{F}_{2^n}\}$ and $G_{F'} = \{(x, F'(x)) | x \in \mathbb{F}_{2^n}\}$ are affine equivalent. It was shown in [6] that EA-equivalence is a particular case of CCZ-equivalence, and furthermore both equivalence relations preserve (up to permutation) the differential table and the extended Walsh spectra. Furthermore, the strength to algebraic cryptanalysis (admittance of low degree I/O equations) is invariant to both equivalence relations. Nevertheless, certain classes of AB (APN) functions derived by applying the CCZ transformation cannot be obtained via classical EA-equivalence [4].

This approach has been successfully used in [3] where the authors conjectured that all functions obtained from an implicitly defined mapping,

$$x^3 + x^2 + x \to x$$

are not EA-equivalent to any power monomial over the field $\mathbb{F}_{2^n}$, for odd $n \geq 3$. Later this statement was corrected not to hold over $\mathbb{F}_{2^3}$ due to the small size of the field [20].

In the rest of this section we propose a recursive method of constructing the APN permutations; a method derived from the implicit mapping $x^3 + x^2 + x \to x$. Note that the Lagrange interpolation for the mapping $x^3 + x^2 + x \to x$ over $\mathbb{F}_{2^3}$ gives the function,

$$g'(x) = x^5 + x^4 + x.$$

This implicit mapping should not be confused with inverse mapping so that $g(x) = (x^3 + x^2 + x)^{-1} \mod x^{2^n} + x$. It is rather a compositional operator that satisfies $g(f(x)) = x$. For $f(x) = x^3 + x^2 + x$ the function $g(x) = x^5 + x^4 + x$ indeed satisfies that $g(f(x)) = x$ which can be verified by computing

$$(x^3 + x^2 + x)^5 + (x^3 + x^2 + x)^4 + (x^3 + x^2 + x) \equiv x \pmod{x^8 + x}.$$

Then if we consider the same implicitly defined mapping $x^3 + x^2 + x \rightarrow x$ over the field $\mathbb{F}_{2^5}$, one computes its explicit form as,

$$g(x) = x^{21} + x^{20} + x^{17} + x^{16} + x^5 + x^4 + x. \tag{9}$$

Then we can deduce that $g(x) = x^{16}(g'(x)+1)+g'(x)$ holds. This however gives a general recursion that relates the function $g' : x^3 + x^2 + x \rightarrow x$ on $\mathbb{F}_{2^{2k-1}}$ to the function $g : x^3 + x^2 + x \rightarrow x$ on $\mathbb{F}_{2^{2k+1}}$, $k \geq 2$, as follows,

$$g(x) = x^{2^{2k}} (g'(x) + 1) + g'(x).$$

It can be verified that the iteration formula generates the same function over $\mathbb{F}_{2^{2k+1}}$ as the function computed using the Lagrange interpolation of implicit mapping $x^3 + x^x + x \rightarrow x$ for all $\mathbb{F}_{2^{2k+1}}$ of practical computational complexity. A rigorous mathematical proof that relates the Lagrange interpolation formula with the simple recursion above is left as an open problem. Nevertheless, we note that the algebraic degree is increased by one in each step of iteration. This is easily verified by noting that $deg_p(g') < 2^{2k-1}$, thus multiplying $g'(x)$ by $x^{2^{2k}}$ will increase the algebraic degree of g' exactly by one. However, we must restrict ourselves to consider polynomials g' with binary coefficients since $\mathbb{F}_{2^{2k-1}} \not\subseteq \mathbb{F}_{2^{2k+1}}$ and therefore g is not well defined.

Remark 2. In accordance with Proposition 1 the function g over $\mathbb{F}_{2^5}$ given by (9) should admit a quadratic dependence between x and y. Indeed, one can verify that $y+g(x) = 0$ gives quadratic algebraic relation (after multiplication with $x^{16}+1$),

$$y(x^{16} + 1) = g(x)(x^{16} + 1) = g'(x)(x + 1) + x + x^{16} = x^{16} + x^6 + x^4 + x^2,$$

where $g'(x) = x^5 + x^4 + x$.

The Lagrange interpolation of an implicit mapping defined by $x^3 + x^2 + x \rightarrow x$ over $\mathbb{F}_{2^7}$ yields the following polynomial,

$$\begin{aligned} g(x) = {} &x^{85} + x^{84} + x^{81} + x^{80} + x^{69} + x^{68} + x^{65} + x^{64} + \\ &+ x^{21} + x^{20} + x^{17} + x^{16} + x^5 + x^4 + x. \end{aligned}$$

This polynomial is of algebraic degree 4 (as $wt(85) = 4$), and it is derived from a cubic function x^3, thus it is a maximally nonlinear APN permutation.

Proposition 2. *Let $g'(x)$ be the compositional inverse mapping of $x \to x^3 + x^2 + x$ over $\mathbb{F}_{2^{2k-1}}$. Then the compositional inverse mapping of the same function over $\mathbb{F}_{2^{2k+1}}$ is defined by,*

$$g(x) = x^{2^{2k}}(g'(x) + 1) + g'(x), \tag{10}$$

and it has the same differential and linear properties as g' and $deg(g)=deg(g')+1$.

This seems to be a peculiar property of the mapping $x \to x^3 + x^2 + x$. In general the recursion above does not guarantee that for a given permutation polynomial g' the polynomial g is also a permutation. Since $deg(g) = \frac{m+1}{2}$ which correspond to the algebraic degree of the inverse coset of Gold mapping $F(x) = x^{2^k+1}$ (with $\gcd(n, k) = 1$) [23], the conjecture on EA-inequivalence of this class of functions in [3] cannot be settled by comparing the algebraic degrees.

Open Problem 1. *Show that the mapping g as defined above is EA-inequivalent to power mappings, in particular to Gold-like mapping $x \to x^3$. In addition, explain why the permutation and APN (AB) property of the input function g' are preserved when the recursion above is applied.*

4   Some Properties of Polynomials in $\mathbb{F}[x, y]/(x^{2^n} + x, y^{2^n} + y)$

As discussed in the previous section it is of interest to consider certain algebraic properties of the functions in the ring of multivariate polynomials $R[x, y] = \mathbb{F}[x, y]/(x^{2^n} + x, y^{2^n} + y)$, where $\mathbb{F} = \mathbb{F}_{2^n}$. Our major concern is the polynomial of the form $y + P(x) = 0$, for which we investigate its *algebraic degree* under certain transformation applied to such a polynomial. That is, for a given function $y = P(x)$, $x, y \in \mathbb{F}_2^n$ we consider a multivariate polynomial in indeterminates x, y in the form $F'(x, y) = y + F(x) = 0$.

Theorem 2. *Let $y + P(x)$ be a polynomial in the ring $R[x, y]$, and let*

$$\min_{S(x) \in \mathbb{F}[x]/x^{2^n}+x} deg\{yS(x) + P(x)S(x)\} = deg\{yQ(x) + P(x)Q(x)\}. \tag{11}$$

Then,

$$deg\{yQ(x) + P(x)Q(x)\} \le deg\{yT(x, y) + P(x)T(x, y)\}, \tag{12}$$

for any $T(x, y) \in R[x, y]$ such that T is strictly a function of both x and y.

Proof. Let $T(x, y)$ denote a polynomial in $R[x, y]$ which minimizes the algebraic degree of $yU(x, y) + P(x)U(x, y)$, $U \in R[x, y]$. Then we may write,

$$T(x, y) = T_0(x) + yT_1(x) + \ldots + y^{2^n - 1}T_{2^n - 1}(x),$$

where $T_i(x) = a_{0,i} + a_{1,i}x + \ldots + a_{2^n-1,i}x^{2^n-1}$, for $a_{j,i} \in \mathbb{F}_{2^n}$. Hence,

$$\begin{aligned} yT(x,y) + P(x)T(x,y) &= y[T_0(x) + yT_1(x) + \ldots + y^{2^n-1}T_{2^n-1}(x)] + \\ &+ P(x)[T_0(x) + yT_1(x) + \ldots + y^{2^n-1}T_{2^n-1}(x)] = \\ &= \underbrace{T_0(x)P(x)}_{K_0(x)} + y\underbrace{[T_0(x) + T_{2^n-1}(x) + T_1(x)P(x)]}_{K_1(x)} + \\ &+ \ldots + y^{2^n-1}\underbrace{[T_{2^n-2}(x) + T_{2^n-1}(x)P(x)]}_{K_{2^n-1}(x)} = \\ &= K_0(x) + yK_1(x) + \ldots + y^{2^n-1}K_{2^n-1}(x). \end{aligned}$$

The terms in the last expansion are distinct, so the algebraic degree of $yT + PT$ is given as the maximum degree of the above terms. Let $y^i K_i(x)$ be the term of the highest degree. Assume now that for any function $Q(x) \in \mathbb{F}[x]/x^{2^n} + x$ it is true that

$$deg\{yQ(x) + P(x)Q(x)\} > deg\{y^i K_i(x)\}.$$

Let now the algebraic degree of $yQ(x) + P(x)Q(x)$ be governed by $yQ(x)$, that is $deg\{yQ(x)\} \geq deg\{P(x)Q(x)\}$. Then we may take $Q(x) := K_i(x)$ which leads to a contradiction unless $i = 0$. When $i = 0$ we get that $K_0(x)$ has a maximum degree among all $y^i K_i(x)$, $i = 0, \ldots, 2^n - 1$. This means that,

$$deg\{K_0(x) + yK_1(x)\} = deg\{K_0(x)\} < deg\{yQ(x) + P(x)Q(x)\} = deg\{yQ(x)\}.$$

This again leads to a contradiction as we may select $Q(x) := K_1(x)$ which gives that,

$$deg\{yK_1(x)\} \leq deg\{K_0(x)\} < deg\{yK_1(x)\}.$$

It remains to consider the case when $deg\{P(x)Q(x)\} \geq deg\{yQ(x)\}$. We again assume that for the highest degree term $y^i K_i(x)$ we have $deg\{P(x)Q(x)\} > deg\{y^i K_i(x)\}$. Now the case $i = 0$ is trivial because assigning $Q(x) := T_0(x)$ gives the equality in the above equation. So, $i > 0$ and then by assumption we have,

$$deg\{K_0(x)\} \leq deg\{y^i K_i(x)\}; \; i > 0.$$

But taking $Q(x) := T_0(x)$ implies that $deg\{P(x)Q(x)\} = deg\{K_0(x)\}$ which leads again to a contradiction. $\qquad\square$

Corollary 2. *Let $y + P(x)$ be a polynomial in $R[x,y]$ such that $deg\{P(x)\} \geq 3$. Then, to check whether there are polynomials of the form $yQ(x,y) + P(x)Q(x,y)$ of algebraic degree less than 3 it is enough to consider Q as being a univariate polynomial in x.*

Moreover, if $yQ(x) + P(x)Q(x)$ is to be of algebraic degree 2, then $Q(x)$ must be affine polynomial of the form,

$$Q(x) = c_0 + a_0 x + a_1 x^{2^1} + \ldots + a_{n-1}x^{2^{n-1}}; \; c_0, a_i \in \mathbb{F}_{2^n}.$$

Proof. The first part of the statement follows directly from Theorem 2. The second part is proved by noting that

$$deg\{yQ(x) + P(x)Q(x)\} = \max\{deg\{yQ(x)\}, deg\{P(x)Q(x)\}\}.$$

Thus for a nonconstant polynomial Q we have $deg\{yQ(x) + P(x)Q(x)\} \geq 2$.   □

Remark 3. Notice that we can find polynomials $T(x,y)$ which gives the same algebraic degree as $Q(x)$ when the equation $(y+P(x))T(x,y)$ is compared to $(y+P(x))Q(x)$ with respect to degree. Hence for finding more independent equations of certain degree we also have to consider the multiplication with $T(x,y)$ as well. A good example is the inverse function $y = x^{-1}$. Then multiplying with x^2 gives $yx^2 + x = 0$ which is of algebraic degree 2, but also multiplication with e.g. x^2y gives $x^2y^2 + xy = 0$, a quadratic equation as well.

A quadratic dependence between x and y may also be obtained through the operation of exponentiating, that is $y^d = P(x)^d$ may give quadratic equations for suitably chosen d. In certain cases the multiplication cannot yield a quadratic dependence while through the powering operation one obtains such a dependence. An excellent example is the Kasami APN power monomial $x^{2^{2k}-2^k+1}$ over the field $\mathbb{F}_{2^n}$, where $gcd(k,n) = 1$, $k > 1$. Then it was verified that $y = x^{2^{2k}-2^k+1}$ does not yield any quadratic equation under the multiplication with some affine polynomial $Q(x)$. On the other hand taking the $(2^k + 1)$th power of the equation $y = x^{2^{2k}-2^k+1}$ will give $y^{2^k+1} = x^{2^{3k}+1}$ which is a quadratic equation in x, y, see [8]. In the next section we combine the two operations of multiplication and squaring to derive a computationally efficient algorithm for computing low degree I/O relations.

4.1   A Fast Algorithm for Finding Low Degree I/O Equations for Sparse Polynomials over $\mathbb{F}_{2^n}$

For the reasons of efficient implementation (table look-up approach) S-box mappings are often chosen to work on a byte level, e.g. the S-boxes of AES are defined as the inverse function in the field $\mathbb{F}_{2^8}$ [12]. An efficient implementation can also be achieved for S-boxes of larger size provided that the defining polynomial $S(x) \in \mathbb{F}_{2^n}[x]$ is efficiently computed ideally both in software and hardware. The choice of large-sized S-boxes has definitely certain advantages over its small-sized counterparts, as cryptographically stronger mappings can be found in larger ambient spaces.

Assuming we are given a sparse polynomial $P(x)$ over $\mathbb{F}_{2^n}$ (the sparseness being induced for the reasons of efficient implementation) we might be interested in the set of all linearly independent low degree I/O equations. Thus the problem is to determine this set in a fast and efficient manner, which is of special importance for large fields. Currently [2], the best known algorithm [13] for

---

[2] Another competitive candidate [1] has a slightly lower time complexity $\mathcal{O}(D^2)$ but a significantly larger memory consumption.

finding low degree annihilators for Boolean function (which is applicable to I/O relations as well) has a time complexity $\mathcal{O}(n2^n D)$ and requires $\mathcal{O}(n2^n)$ memory, where $D = \binom{n}{d}$. For instance if $n = 32$, $d = 3$, the time and memory complexity are 2^{49} and 2^{37} respectively.

For simplicity, the algorithm that follows considers only the existence of quadratic I/O equations. The concept is easily generalized for any small pre-specified value $2 \leq d \ll n$. Note that for any polynomial $y + P(x) = 0$ over the field $\mathbb{F}_{2^n}$ the quadratic relationship in x and y, if it exists, can be written as,

$$\alpha + \sum_i a_i x^{2^i} + \sum_j b_i y^{2^i} + \sum_{i,j} c_{i,j} y^{2^i + 2^j} + \sum_{i,j} d_{i,j} x^{2^i + 2^j} + \sum_{i,j} e_{i,j} x^{2^i} y^{2^j} = 0, \quad (13)$$

where $0 \leq i, j \leq n - 1$ and $a_i, \ldots e_{i,j}, \alpha \in \mathbb{F}_{2^n}$.

Let $y = P(x)$ be a given polynomial (not necessarily permutation) over the field $\mathbb{F}_{2^n}$. Throughout this section we let $Q(x)$ denote a quadratic polynomial in x over $\mathbb{F}_2$, that is, $Q(x) = \sum_i \alpha_i x^{d_i}$, where $\alpha_i \in \mathbb{F}_{2^n}$ and $wt(d_i) \leq 2$. Also, $NQ(x)$ will denote a polynomial whose monomials are of degree greater than 2. Then we introduce the degree ordering so that all terms of any polynomial satisfy the following rule. The term $\alpha_i x^{d_i}$ comes before the term $\alpha_j x^{d_j}$ if $wt(d_i) > wt(d_j)$. In the case that $wt(d_i) = wt(d_j)$ then the deciding rule is $d_i > d_j$. For instance $P(x) = x^{12} + x^{11} + x^8 + x^3$ is then written as $P(x) = x^{11} + x^{12} + x^3 + x^8$, and accordingly $NQ(x) = x^{11}$, $Q(x) = x^{12} + x^3 + x^8$. Let us consider the set of polynomials

$$\{y^d = P(x)^d; \ 1 \leq d \leq 2^n - 1, \ 1 \leq wt(d) \leq 2\},$$

for all of which we may write $y^d = NQ_{(d)}(x) + Q_{(d)}(x)$. Then if $NQ_{(d)}(x) = 0$ the polynomial $y + Q(x) \in R[x, y]$ gives rise to quadratic equations over $\mathbb{F}_2$. When $NQ_{(d)}(x) \neq 0$ then we express the leading term αx^u of $NQ_{(d)}(x)$ as,

$$x^u = \alpha^{-1}(y^d + NQ'_{(d)}(x) + Q_{(d)}(x)),$$

where $NQ'_{(d)}(x) = NQ_{(d)}(x) - \alpha x^u$. This gives $n + n(n-1)/2$ equations (for all d s.t. $1 \leq wt(d) \leq 2$) that express the higher degree terms via lower degree polynomials. The only thing we should be concerned about is not to replace the leading monomial if it has already been replaced in some previous step. We proceed in the same manner by computing $y^{2^i} x^{2^j} = P(x)^{2^i} x^{2^j}$ which gives n^2 new equations. Hence in total we get $n + n(n-1)/2 + n^2$ equations that replace the high degree terms with low degree terms.

This approach is in accordance with the general expression given by (13). That is, the operations of powering and multiplication by x^{2^i} will give the terms of the following forms y^{2^i}, $y^{2^i + 2^j}$, $x^{2^i} y^{2^j}$ which covers the three sums in (13), whereas the quadratic functions $Q_i(x)$ stands for the remaining two sums. Also, note that we need not take into account the equations of the form $y^d(x)A(x) = P^d(x)A(x)$ as they only give linearly dependent equations of the form,

$$(y(x)^d + P(x)^d)a_0 + (y(x)^d + P(x)^d)a_1 x^{2^0} + \cdots + (y(x)^d + P(x)^d)a_n x^{2^{n-1}} = 0.$$

Input: An arbitrary function $P : \mathbb{F}_{2^n} \to \mathbb{F}_{2^n}$.
1. For $i = 0, \ldots, n-1$ compute

$$y^{2^i} = P(x)^{2^i} = NQ_{(i)}(x) + Q_{(i)}(x)$$

Sort and express the highest degree term x^u as:

$$x^u = \alpha^{-1}(y^{2^i} + NQ'_{(i)}(x) + Q_{(i)}(x)); \quad NQ'_{(i)}(x) = NQ_{(i)}(x) - \alpha x^u.$$

2. For $0 \le i, j \le n-1$, $i \ne j$ compute:

$$y^{2^j + 2^i} = P(x)^{2^j} \cdot P(x)^{2^i} = NQ_{(j,i)}(x) + Q_{(j,i)}(x),$$

Sort and express the highest degree term x^u in $NQ_{(j,i)}(x)$ as above
Express, if possible, high degree monomials (including x_u) with $Q_k(x, y)$
If all nonquadratic terms may be expressed with some $Q_{(j,i)}(x, y)$ output
quadratic function $Q(x, y) = 0$.
3. For $0 \le j, k \le n-1$, compute:

$$y^{2^j} x^{2^k} = P(x)^{2^j} x^{2^k} = NQ_{(j,k)}(x) + Q_{(j,k)}(x).$$

Repeat the same computation as in the step 2.

Fig. 1. Algorithm for finding quadratic equations over a field

Below we depict the formal steps of the algorithm. Due to the lack of space a small example describing how the algorithm works is given in the Appendix. Assuming that $P(x)$ is sparse, containing $k \ll 2^n$ terms, the time complexity of our algorithm is as follows. For convenience, we assume that the input to the algorithm is a polynoimal $P(x)$ with binary coefficients. In this way we substantially simplify the analysis leaving the complexity estimate for the arbitrary form of P to the extended version of this paper. However, the polynomials with binary coefficents are of particular interest for efficient implementations.

Each exponentiation of the form $P(x)^{2^i}$ takes kn operations, and in addition k operations for sorting. The exponentiation estimate is justified by noting that computing $(x^{e_j})^{2^i} = x^{e_j 2^i} \bmod x^{2^n} + x$ (for any monomial $x^{e_j} \in P(x)$) corresponds to computing $e_j 2^i \bmod 2^n - 1$. Representing e_j as a binary n-bit integer this modular multiplication with 2^i is equivalent to a left circular shift of e_j with i bit positions. For instance, computing $e_j 2^i = 7 \cdot 8 \equiv 25 \pmod{31}$ over $\mathbb{F}_{2^5}$ is performed as $e_j = (0, 0, 1, 1, 1) \to (1, 1, 0, 0, 1)$. Therefore, computing $P(x)^{2^i}$ for $i = 0, \ldots, n-1$ takes $n^2 k$ operations. Similarly, $P(x)^{2^i + 2^j} = P(x)^{2^i} \cdot P(x)^{2^j}$ takes $k^2 n$ operations of simple modulo $2^n - 1$ additions of exponents, and k^2 for sorting. Thus, in total $\binom{n}{2}(k^2 n + k^2)$ operations are performed. Finally, computing n^2 multiplications $P(x)^{2^i} x^{2^j}$, $i, j = 0, \ldots, n-1$, requires $n^2 k$ modular additions. The complexity of this step is therefore $n^3 k$, and in addition $n^2 k$ operations are needed for sorting.

Thus, for $d = 2$ the time complexity T_C is dominated by the second and third step of algorithm,

$$T_C \approx \mathcal{O}(n^3 k^2).$$

This is a polynomial running time algorithm for small k, and for instance if $k \approx n$ then $T_C = \mathcal{O}(n^5)$. On the other hand, for a random polynomial $P(x) \in \mathbb{F}_{2^n}[x]$ the expected value of k is 2^{n-1} so that $T_C = \mathcal{O}(n^3 2^{2n})$, which is larger than $\mathcal{O}(n2^n D)$, $D = \binom{n}{2}$. But these classes of functions, containing many monomials, are definitely of no interest in cryptographic applications.

A generalization for finding I/O equations of degree $d > 2$ is straightforward. In this case the term dominating the total complexity is the evaluation of the terms $y^{2^{j_1}} y^{2^{j_2}} \cdots y^{2^{j_d}}$ as k^d multiplications are needed. The complexity of our algorithm is then well approximated by $T_c \approx \mathcal{O}(n^{d+1} k^d)$, which is still much smaller compared to $\mathcal{O}(n^{d+1} 2^n)$ (using $D = \binom{n}{d} \approx n^d$) for relatively small d.

5   Conclusions

In this paper we have addressed some important issues related to the properties of cryptographically significant mappings over finite fields. It would be of interest to get a deeper unerstanding for the recursive procedure of APN permutations given in Section 3.1, and to further investigate the possibility of other recursive methods in this context.

References

1. Armknecht, F., Carlet, C., Gaborit, P., Künzli, S., Meier, W., Ruatta, O.: Efficient computation of algebraic immunity for algebraic and fast algebraic attacks. In: Vaudenay, S. (ed.) EUROCRYPT 2006. LNCS, vol. 4004, pp. 147–164. Springer, Heidelberg (2006)
2. Biham, E., Shamir, A.: Differential cryptanalysis of DES-like cryptosystems. Journal of Cryptology 4(1), 3–72 (1991)
3. Breveglieri, L., Cherubini, A., Macchetti, M.: On the generalized linear equivalence of functions over finite fields. In: Lee, P.J. (ed.) ASIACRYPT 2004. LNCS, vol. 3329, pp. 79–91. Springer, Heidelberg (2004)
4. Budaghyan, L.: The simplest method for constructing APN polynomials EA-inequivalent to power functions. In: Carlet, C., Sunar, B. (eds.) WAIFI 2007. LNCS, vol. 4547, pp. 177–188. Springer, Heidelberg (2007)
5. Budaghyan, L., Carlet, C., Pott, A.: New classes of almost bent and almost perfect nonlinear polynomials. IEEE Trans. on Inform. Theory IT-52(3), 1141–1152 (2006)
6. Carlet, C., Charpin, P., Zinoviev, V.: Codes, bent functions and permutations suitable for DES-like cryptosystems. Designs, Codes and Cryptography 15(2), 125–156 (1998)
7. Chabaud, F., Vaudenay, S.: Links between differential and linear cryptanalysis. In: De Santis, A. (ed.) EUROCRYPT 1994. LNCS, vol. 950, pp. 356–365. Springer, Heidelberg (1995)
8. Cheon, J.H., Lee, D.H.: Resistance of S-boxes against algebraic attacks. In: Roy, B., Meier, W. (eds.) FSE 2004. LNCS, vol. 3017, pp. 83–94. Springer, Heidelberg (2004)

9. Courtois, N.: Higher order correlation attacks, XL algorithm and cryptanalysis of Toyocrypt. In: Lee, P.J., Lim, C.H. (eds.) ICISC 2002. LNCS, vol. 2587, pp. 182–199. Springer, Heidelberg (2003)

10. Courtois, N., Debraize, B., Garrido, E.: On exact algebraic [non-]immunity of S-boxes based on power functions. In: Batten, L.M., Safavi-Naini, R. (eds.) ACISP 2006. LNCS, vol. 4058, pp. 76–86. Springer, Heidelberg (2006)

11. Courtois, N., Pieprzyk, J.: Cryptanalysis of block ciphers with overdefined systems of equations. In: Zheng, Y. (ed.) ASIACRYPT 2002. LNCS, vol. 2501, pp. 267–287. Springer, Heidelberg (2002)

12. Daemen, J., Rijmen, V.: The Design of Rijndael. Springer, Berlin (2002)

13. Didier, F.: Using Wiedemann's algorithm to compute the immunity against algebraic and fast algebraic attacks. In: Barua, R., Lange, T. (eds.) INDOCRYPT 2006. LNCS, vol. 4329, pp. 236–250. Springer, Heidelberg (2006)

14. Dobbertin, H.: Almost perfect nonlinear power functions on $GF(2^n)$: The Welch case. IEEE Trans. on Inform. Theory IT-45(4), 1271–1275 (1999)

15. Dobbertin, H.: Almost perfect nonlinear power functions over $GF(2^n)$: The Niho case. Inform. Comput. 151, 57–72 (1999)

16. Faugère, J.-C.: A new efficient algorithm for computing Gröbner basis without reduction to 0 F_5. In: Proceedings of ISSAC 2002, pp. 75–83. ACM Press, New York (2002)

17. Fraenkel, S.A., Yesha, Y.: Complexity of problems in games, graphs, and algebraic equations. Discr. Appl. Math. 1, 15–30 (1979)

18. Hou, X.D.: Affinity of permutations of $\mathbb{F}_{2^n}$. Discr. Appl. Math. vol. 154(2), 313–325 (2006)

19. Knudsen, L.R.: Quadratic relations in Khazad and Whirlpool. NESSIE report NES/DOC/UIB/WP5/017/1 (2002)

20. Macchetti, M.: Addendum to On the generalized linear equivalence of functions over finite fields. Cryptology ePrint Archive, Report2004/347 (2004), http://eprint.iacr.org/

21. Matsui, M.: Linear cryptanalysis method for DES cipher. In: Helleseth, T. (ed.) EUROCRYPT 1993. LNCS, vol. 765, pp. 386–397. Springer, Heidelberg (1994)

22. Murphy, S., Robshaw, M.: Essential algebraic structure within the AES. In: Yung, M. (ed.) CRYPTO 2002. LNCS, vol. 2442, pp. 1–16. Springer, Heidelberg (2002)

23. Nyberg, K.: Differentially uniform mappings for cryptography. In: Helleseth, T. (ed.) EUROCRYPT 1993. LNCS, vol. 765, pp. 55–64. Springer, Heidelberg (1994)

24. Shamir, A., Patarin, J., Courtois, N., Klimov, A.: Efficient algorithms for solving overdefined systems of multivariate polynomial equations. In: Preneel, B. (ed.) EUROCRYPT 2000. LNCS, vol. 1807, pp. 392–407. Springer, Heidelberg (2000)

Appendix

Example 2. *Let us again consider the function $y = g(x)$ over $\mathbb{F}_{2^5}$ given by,*

$$g(x) = x^{21} + x^{20} + x^{17} + x^{16} + x^5 + x^4 + x,$$

which is of algebraic degree $d = 3$. As we already remarked

$$y(x^{16} + 1) = g(x)(x^{16} + 1) = g'(x)(x + 1) + x + x^{16} = x^{16} + x^6 + x^4 + x^2,$$

where $g'(x) = x^5 + x^4 + x$. *This quadratic algebraic equation was obtained using the special form of* g. *Of course there might be other quadratic equations and we check for these applying our algorithm; these will be indicated by* $*$

$$y = x^{21} + x^{20} + x^{17} + x^5 + x^{16} + x^4 + x; \Rightarrow x^{21} = y + Q_0(x),$$

$$y^2 = x^{11} + x^{10} + x^9 + x^3 + x^8 + x^2 + x; \Rightarrow x^{11} = y^2 + Q_1(x),$$

$$y^4 = x^{22} + x^{20} + x^{18} + x^6 + x^{16} + x^4 + x^2; \Rightarrow x^{22} = y^4 + Q_2(x),$$

$$y^8 = x^{13} + x^{12} + x^9 + x^5 + x^8 + x^4 + x; \Rightarrow x^{13} = y^8 + Q_3(x),$$

$$y^{16} = x^{26} + x^{24} + x^{18} + x^{10} + x^{16} + x^8 + x^2; \Rightarrow x^{26} = y^{16} + Q_4(x),$$

$*$ $$y^3 = x^{21} + x^{11} + Q_5(x); \Rightarrow y^3 = y + Q_0(x) + y^2 + Q_1(x) + Q_5(x),$$

$*$ $$y^5 = x^{22} + x^{21} + Q_6(x); \Rightarrow y^5 = y^4 + Q_2(x) + y + Q_0(x) + Q_6(x),$$

$*$ $$y^6 = x^{22} + x^{11} + Q_7(x); \Rightarrow y^6 = y^4 + Q_2(x) + y^2 + Q_1(x) + Q_7(x),$$

$*$ $$y^9 = x^{21} + x^{13} + Q_8(x); \Rightarrow y^9 = y + Q_0(x) + y^8 + Q_3(x) + Q_8(x),$$

$*$ $$y^{10} = x^{13} + x^{11} + Q_9(x); \Rightarrow y^{10} = y^8 + Q_3(x) + y^2 + Q_1(x) + Q_8(x),$$

$*$ $$y^{12} = x^{22} + x^{13} + Q_{10}(x); \Rightarrow y^{12} = y^4 + Q_2(x) + y^8 + Q_3(x) + Q_{10}(x),$$

$*$ $$y^{17} = x^{26} + x^{21} + Q_{11}(x); \Rightarrow y^{17} = y^{16} + Q_4(x) + y + Q_0(x) + Q_{11}(x),$$

$*$ $$y^{18} = x^{26} + x^{11} + Q_{12}(x); \Rightarrow y^{18} = y^{16} + Q_4(x) + y^2 + Q_1(x) + Q_{12}(x),$$

$*$ $$y^{20} = x^{26} + x^{22} + Q_{13}(x); \Rightarrow y^{20} = y^{16} + Q_4(x) + y^4 + Q_2(x) + Q_{13}(x),$$

$*$ $$y^{24} = x^{26} + x^{13} + Q_{14}(x); \Rightarrow y^{24} = y^{16} + Q_4(x) + y^8 + Q_3(x) + Q_{14}(x),$$

$*$ $$yx = x^{22} + x^{21} + Q_{15}(x); \Rightarrow yx = y^4 + Q_2(x) + y + Q_0(x) + Q_{15}(x),$$

$$\vdots$$

Remark that the maximum number of linearly independent quadratic equations over $\mathbb{F}_2$ for $n = 5$ is $\binom{10}{2} = 45$, it is the number of terms of the form $x_i x_j, y_i y_j, x_i y_j$. Exceeding this number would imply the existence of linear I/O equations which is not possible.

Author Index